G Proteins
and
Calcium Signaling

Editor

Paul H. Naccache

Associate Professor
Inflammation and Immunology-Rheumatology Research Unit
Laval University
Ste. Foy, Québec

CRC Press, Inc.
Boca Raton, Florida

Library of Congress Cataloging-in-Publication Data

G proteins and calcium signaling/editor, Paul Naccache.
    p.   cm.
   Includes bibliographies and index.
   ISBN 0-8493-4572-3
   1. G proteins--Physiological effect.   2. Calcium--Physiological
effect.   3. Cellular control mechanisms.   I. Naccache, Paul.
   [DNLM: 1. Calcium Channels--physiology.   2. Guanine Nucleotide
Regulatory Protein--physiology.   3. Phospholipases--metabolism.
4. Signal Transduction.   QU 55 G111]
QP552.G16P76 1990
574.87'5--dc19
DNLM/DLC
for Library of Congress
                                                                                    89-582
                                                                                    CIP

    Direct all inquiries to CRC Press, Inc., 2000 Corporate Blvd., N.W., Boca Raton, Florida, 33431.

© 1990 by CRC Press, Inc.

International Standard Book Number 0-8493-4572-3

Library of Congress Card Number 89-582
Printed in the United States

PREFACE

Whereas much information has been available for several years concerning the "early" (the binding sites, their nature, and characteristics) and late (the second messengers such as calcium and more recently Insl,4,5P$_3$ and diglycerides) events of the excitation-response coupling sequence in nonmuscle cells,[1] it is only within the last couple of years that similarly detailed data has been obtained about the molecular nature of the elements linking these events together.

The articles in this book are meant to document, by means of selected examples, this recent breakthrough. The articles chosen focus on the concept that guanine nucleotide binding proteins (G proteins) act as coupling agents not only in cyclic nucleotide[2] but also in calcium-dependent systems. This is a major development in cellular physiology, the dissemination of the implication of which is the goal of this book. Special care was taken to illustrate the roles of the G proteins during varied physiological activation processes as well as their potential implications in the mechanisms of oncogenic transformation.

An overview of the mechanism of action of the bacterial toxin from *Bordetella pertussis* and its use as a tool in the investigation of the potential roles of G proteins (Chapter 1), is followed by a discussion of the involvement of the G proteins in the mediation of the activation of two major effector enzyme systems in calcium-dependent systems, namely phospholipases C and A$_2$ (Chapters 3 and 4). Chapters 5 and 6 detail the presently available evidence linking G proteins to the voltage-dependent gating of calcium movements across the plasma membrane and to intracellular calcium homeostasis and its relationship to the Insl,4,5P$_3$-sensitive mobilization of calcium. The generality of the role of G proteins in signal transduction is further illustrated by the description of their involvement in sensory systems (Chapters 7 and 8). This short collection also includes a review of the relationship between G proteins and oncogenic products implicated in signal transduction (Chapter 8).

This overview presents therefore the major elements of the presently understood roles of G proteins in calcium signaling. The present research emphasis on the cloning and sequencing of the G proteins ensures that the subsequent volumes dedicated to this field will present more detailed molecular analyses of the interactions between surface binding sites (receptors), the G proteins, and the effector systems (phospholipase C, A$_2$, or other as yet undefined systems). In the meantime, it is hoped that this volume will help expose a wider audience to this major development in cellular physiology and stimulate the investigation of its ramifications, and limitations, in as large a variety of experimental situations as possible.

REFERENCES

1. **Putney, J. W., Jr.,** *Phosphoinositides and Receptor Mechanisms,* Alan R. Liss, New York, 1986.
2. **Gilman, A. G.,** Guanine nucleotide binding regulatory proteins and dual control of adenylate cyclase, *Cell,* 36, 577, 1984.

THE EDITOR

Paul H. Naccache, Ph.D. is professeur adjoint (subventionel) in the Department of Medicine of Laval University in Québec City. Previously, he held a faculty position at the University of Connecticut Health Center in Farmington, Connecticut.

His educational background includes a B.S. in physics and an M.S. in physiology from the American University of Beirut, Beirut, Lebanon, and a Ph.D. in physiology from the University of Connecticut Health Center before joining the faculty there.

Dr. Naccache has published extensively in cellular physiology, especially as related to neutrophils. He is a member of several professional societies (American Physiological Society, the New York Academy of Sciences, and the American Society for Cell Biology), and holds fellowships from the Fonds de la Recherche en Santé du Québec and the Arthritis Society.

CONTRIBUTORS

Joan Heller Brown
Department of Pharmacology
University of California
San Diego School of Medicine
La Jolla, California

Richard C. Bruch
Department of Neurobiology and
  Physiology
Northwestern University
Evanston, Illinois

Ronald M. Burch
Pain, Inflammation and Bradykinin
  Antagonist Program
Nova Pharmaceutical Corporation
Baltimore, Maryland

Alan Fein
Department of Physiology
University of Connecticut Health Center
Farmington, Connecticut

Tarun K. Ghosh
Department of Biochemistry
University of Maryland
Baltimore, Maryland

Donald L. Gill
Department of Biochemistry
University of Maryland
Baltimore, Maryland

Alan Hall
Institute of Cancer Research
Chester Beatty Laboratories
London, England

Jürgen Hescheler
Physiologisches Institut
Universität des Saarlandes
Hamburg/Saar, West Germany

Julienne M. Mullaney
Department of Biochemistry
University of Maryland
Baltimore, Maryland

Richard Payne
Department of Zoology
University of Maryland
College Park, Maryland

Lawrence A. Quilliam
Department of Pharmacology
University of California
San Diego School of Medicine
La Jolla, California

Walter Rosenthal
Institut für Pharmakologie
Freien Universität Berlin
Berlin, West Germany

Michio Ui
Department of Physiological Chemistry
Faculty of Pharmaceutical Sciences
University of Tokyo
Tokyo, Japan

Fahmy I. Tarazi
Department of Biochemistry
University of Maryland
Baltimore, Maryland

TABLE OF CONTENTS

NATURE OF G PROTEINS AND RELATIONSHIP TO ONCOGENES AND CELL GROWTH

Chapter 1
G Proteins Identified as Pertussis Toxin Substrates......................................3
M. Ui

Chapter 2
Oncogene Products Involved in Signal Transmission....................................29
A. Hall

REGULATION OF PHOSPHOLIPASE A$_2$ AND C

Chapter 3
Regulation of Phospholipase A$_2$ by G Proteins...47
R. M. Burch

Chapter 4
The Involvement of a GTP-Binding Regulatory Protein in Muscarinic-Stimulated Inositol Phospholipid Metabolism...59
L. A. Quilliam and J. H. Brown

G PROTEINS IN PHOSPHOLIPASE C-INDEPENDENT CALCIUM MOVEMENTS

Chapter 5
G Protein Control of Voltage-Dependent Ca^{2+} Currents...............................79
J. Hescheler and W. Rosenthal

Chapter 6
Inositol Phosphate- and Guanine Nucleotide-Activated Calcium Translocation within Cells...95
D. L. Gill, J. M. Mullaney, T. K. Ghosh, S. E. Devine, and M. Yu

G PROTEINS IN SENSORY TRANSDUCTION

Chapter 7
G Proteins in Olfactory Neurons...123
R. C. Bruch

Chapter 8
The Release of Calcium by Light in the Photoreceptors of Invertebrates...............135
R. Payne and A. Fein

INDEX...157

Nature of G Proteins and Relationship to Oncogenes and Cell Growth

Chapter 1

G PROTEINS IDENTIFIED AS PERTUSSIS TOXIN SUBSTRATES

Michio Ui

TABLE OF CONTENTS

I.   Pertussis Toxin as a Probe of G Proteins 4
      A.   Abolition of Epinephrine-Induced Hyperglycemia by Pertussis
            Vaccine — The First Clue to a Probe for G Proteins..................... 4
      B.   A G Protein Involved in Adenylate Cyclase Inhibition
            Identified as the Target of Pertussis Toxin 4
      C.   A-B Structure of Pertussis Toxin Essential for the G Protein
            Probe.. 5

II.   G Proteins Serving as the Substrate of Pertussis
      Toxin-Catalyzed ADP-Ribosylation 7
      A.   Diversity of G Proteins.. 7
      B.   G Proteins as Transducers between Receptors and Effectors 9
      C.   Uncoupling of G Proteins from Receptors after being ADP-
            Ribosylated by Pertussis Toxin 10

III. Pertussis Toxin Substrate G Proteins Mediating Signaling Other than
      the Adenylate Cyclase System 11
      A.   Pertussis Toxin-Susceptible Phospholipase C Activation 11
      B.   Pertussis Toxin-Susceptible Phospholipase A$_2$ Activation 12
      C.   Pertussis Toxin-Susceptible Potassium Channel Activation 13
      D.   Pertussis Toxin-Susceptible Calcium Channel Regulation 13
      E.   Pertussis Toxin-Susceptible Cell Proliferation and
            Differentiation 14

IV.   Conclusion... 17

Acknowledgments 18

References... 18

I. PERTUSSIS TOXIN AS A PROBE OF G PROTEINS

A. ABOLITION OF EPINEPHRINE-INDUCED HYPERGLYCEMIA BY PERTUSSIS VACCINE — THE FIRST CLUE TO A PROBE FOR G PROTEINS

Pertussis toxin is now widely used as a unique and valuable probe for G proteins (the guanine nucleotide-binding regulatory proteins). The original impetus for the introduction of pertussis toxin into this field of research came from the pioneering work of Ui and colleagues as to how a prior single-time injection of rats with pertussis vaccine prevents hyperglycemic responses of these animals to the subsequent challenge with epinephrine, an agonist of the α- and β-adrenergic receptors.[1-5] Secretion of insulin from the pancreatic islet B-cells is under the adrenergic control; the stimulation is caused via the β-, while the inhibition is caused via the α_2-adrenergic receptors. The inhibition of the hormone secretion via the α_2-adrenergic receptors is thus a cause of the catecholamine-induced hyperglycemia in normal rats. In rats treated with pertussis vaccine, however, epinephrine acted as such a potent β-adrenergic agonist that much insulin was released into the blood circulation. A large amount of insulin thus secreted interferes with hyperglycemia otherwise developing in response to the catecholamine. Pertussis vaccine was thus expected to contain a novel factor that modulates adrenergic responses of mammals by converting those of an α-type to those of a β-type.

This factor was purified from the 2-d culture supernatant of *Bordetella pertussis* and proved to display just the same effect as pertussis vaccine when it was injected into rats at a dose as low as 1 μ/100 g of body weight.[6,7] The effect was characterized by its long duration; epinephrine failed to cause hyperglycemia for several weeks in the rats once injected with the purified factor, being suggestive of irreversible modification of its target(s) in the body. The factor was a protein with a molecular weight (M_r) of 117,000, and termed islet-activating protein (or IAP as the abbreviated term). IAP was soon identified as the same entity as pertussis toxin, an exotoxin that, being produced by *B. pertussis,* plays a role in aiding the organism establish and maintain an infection. This toxin is referred to as either IAP or pertussis toxin in this chapter.

Pertussis toxin thus purified was effective on intact cell preparations *in vitro*, although no effect developed during a definite lag period of 1 to 3 h.[8-11] A search was made for the target of IAP when it interacted with intact cells. Rat pancreatic islets were incubated in a glucose-containing medium for several hours. The inclusion of epinephrine in the medium inhibited insulin secretion from the cells as a result of stimulation of the α_2-adrenergic receptors. When the catecholamine-containing medium was further supplemented with a low concentration of IAP, however, the inhibition was reversed and insulin started to accumulate in the medium after a lag time of 1 h. Changes in the rate of insulin secretion thus induced by various concentrations of epinephrine were well correlated with the same-directional changes in the intracellular concentration of cAMP. Conceivably, cAMP is a critical factor in triggering insulin secretion as the second messenger in islet cells. The catecholamine-induced inhibition of the hormone secretion results from α_2-adrenergic inhibition of adenylate cyclase, the enzyme producing cAMP. The degree of the epinephrine-induced inhibition was reduced by IAP similarly for both insulin secretion and cAMP production in the cells. The target of IAP was hence considered to be the receptor-adenylate cyclase coupling, as was later confirmed in studies with the membrane preparation from islets.[12]

B. A G PROTEIN INVOLVED IN ADENYLATE CYCLASE INHIBITION AS THE TARGET OF PERTUSSIS TOXIN

Pertussis toxin interacts with various types of cells as well as islet cells. All the receptors found to be involved in adenylate cyclase inhibition served as the target of IAP in these toxin-susceptible cells. They include muscarinic acetylcholine and A_1-adenosine receptors in rat heart cells,[13] α_2-adrenergic, muscarinic, and opiate receptors in NG108-15 cells,[14] and α_2-adrenergic and A_1-adenosine receptors in rat and hamster adipocytes.[15,16] Stimulation of these receptors by specific agonists caused decreases in cAMP in intact cells or inhibition of adenylate cyclase in membranes prepared therefrom. The decreases and inhibition were reversed effectively by prior exposure of the cells to IAP for 1 to 3 h. The adenylate cyclase activity was inhibited by GTP as well under certain conditions in membranes from 3T3 fibroblasts.[17] GTP was not inhibitory in membranes that were prepared from IAP-treated cells. The GTP-binding site, rather than the receptor site, in the receptor-adenylate cyclase system, appeared to be a real target of pertussis toxin.

Katada and Ui were the first who showed that IAP was as effective in cell-free membrane preparations as in intact cells, provided the cell-free preparations were fortified with NAD and ATP.[18,19] NAD was essential because it served as one of the substrates of the reaction catalyzed by IAP, while ATP stimulated the catalytic activity of the toxin. Incubation of membranes from rat C6 glioma cells with IAP, ATP, and radioactive NAD, followed by electrophoretic analysis of the labeled membrane proteins on sodium dodecyl sulfate poly-acrylamide gels revealed that IAP catalyzed the transfer of the ADP-ribosyl moiety of NAD into a membrane protein with $M_r = 41{,}000$ very selectively. In another series of experiments, membranes similarly labeled with radioactive NAD in the presence of IAP were subjected to mild digestion by trypsin before the separation of labeled proteins on electrophoretic gels. The digestion pattern of the $M_r = 41{,}000$ protein serving as the substrate of IAP-catalyzed ADP-ribosylation was profoundly affected by the inclusion of GTPγS, a nonhydrolyzable GTP analog, or NaF in the digestion medium. Thus, the IAP substrate proved to be one of the membrane GTP-binding proteins to which GTP or NaF binds specifically.[19]

The additional important findings reported by Katada and Ui in 1982 were related to evidence for the ideas that ADP-ribosylation of IAP-substrate G protein occurs in intact cells as well when the cells are exposed to the toxin and that this substrate protein is involved in receptor-mediated inhibition of adenylate cyclase. For this purpose, they cultured a series of dishes of C6 cells containing various concentrations of IAP for 3 h and prepared membranes from each of the dishes of cells. A batch of these membrane preparations was ADP-ribo-sylated by [α-^{32}P]NAD with further addition of a saturating concentration of the toxin, while the other batch was studied for GTP-dependent inhibition of adenylate cyclase activity. The membrane adenylate cyclase activity increased progressively, as a result of reversal of the inhibition, and the incorporation of ^{32}P into the membrane $M_r = 41{,}000$ protein decreased reciprocally, as the concentration of IAP in the culture was increased.[19] The toxin concentration-dependent plot of adenylate cyclase activities thus obtained was an exact mirror image of the similar plot of ADP-ribosylation of the membrane protein reflecting the residual amounts of the same protein that had not been ADP-ribosylated during the culture with these varying concentrations of the toxin. These results afforded convincing evidence for their proposal that a G protein involved in adenylate cyclase inhibition is selectively ADP-ribo-sylated during exposure of cells to IAP and that the protein is incapable of inhibiting the cyclase any longer after being thus modified. G_i (or N_i), the presence of which was first predicted by Rodbell,[20] was thus identified as a specific target protein of IAP.

C. A-B STRUCTURE OF PERTUSSIS TOXIN ESSENTIAL FOR THE G PROTEIN PROBE

One of the great advantages of IAP as a G protein probe is that the toxin is capable of modification of the target G protein in intact cells without causing any damage to other cellular functions. It is because the big toxin molecule with $M_r = 117{,}000$ enters the intact cells by an exquisite means that depends on the unique subunit structure of the toxin.

Pertussis toxin is a hexameric protein consisting of five different subunits named S_1,

S_2, S_3, S_4, and S_5 on the basis of the order of their M_r values.[21] These subunits assemble to form a holotoxin molecule by noncovalent binding. The biggest subunit, S_1, was termed an A(Active)-protomer, since it is an active enzyme catalyzing ADP-ribosylation; the subunit displays the enzymic activity when it is resolved from the residual pentamer and the intra-peptide disulfide bonds are reductively cleaved.[22] ATP binds to the S_3 subunit and then exerts its allosteric effect to accelerate resolution of the A-protomer from the pentamer (unpublished observation). The pentamer is of such a complicated structure that two dimers, Dimer-1 composed of S_2 and S_4 and Dimer-2 composed of S_3 and S_4, are connected with each other by means of the smallest subunit, S_5. The pentamer is termed a B(Binding)-oligomer, since it binds to particular glycoproteins on the surface of cells via Dimer-1 (or Dimer-2 with a lower affinity). This binding enables the holotoxin molecules to enter the cells as a result of physiological internalization of the membrane glycoproteins to which they have bound.[23] The time required for the internalization of membrane proteins reflects the lag time necessarily preceding the onset of actions of IAP on intact cells (see Section I.A).

The A-protomer is then released from the holotoxin thus internalized into intact cells owing to the high intracellular concentration of ATP and further subjected to reductive cleavage of the intramolecular disulfide bonds as a result of its interaction with cytosolic reduced glutathione,[22] thereby yielding an ADP-ribosyltransferase which substrates are G proteins including G_i, as described in Section I.B, and other similar proteins, as will be described in Section II.A in this chapter. Thus, the entity directly responsible for the mod-ification of G proteins inside the cell membrane is the A-protomer moiety, an enzyme that is *injected* into the cell through the B-oligomer moiety. The product of the enzymic reaction, the ADP-ribosylated cysteine residue in the substrate protein molecule (see Section II.A), is stable in the cell. Only a few moles of IAP can modify, therefore, a large number of its target molecules, if the cell is exposed to the toxin for a long time.[24]

Prior activation of IAP, the procedure to yield the active A-protomer from the holotoxin, is a prerequisite for the toxin target G proteins to be ADP-ribosylated by added NAD in the absence of cellular structure or components or in their partially or totally purified forms. A simple procedure advisable is an incubation of the holotoxin with 2 to 5 mM dithiothreitol and ATP for 10 to 20 min prior to application to the ADP-ribosylation mixture. Much (usually two or three orders of magnitude) higher concentrations of the toxin are needed for experiments in a cell-free system than those applicable to intact cells, since the time for ADP-ribosylation has to be kept much shorter owing to instability of the activated A-protomer or rapid decomposition of NAD by particulate-bound and cytosolic hydrolases under the former conditions.

ADP-ribosylation of G proteins is not the only effect of IAP on mammalian cells. The binding of the B-oligomer to the cell surface glycoproteins via both of the two dimers, unlike the *monovalent* binding via Dimer-1, causes crosslinking of these membrane proteins in T lymphocytes, thereby triggering intracellular signaling that probably arises from phospho-lipase C activation and that eventually leads to DNA synthesis and cell proliferation.[23,25,26] Pertussis toxin is thus a T-cell mitogen. Signals mobilizing Ca^{2+} could be inhibited by IAP owing to ADP-ribosylation of involved G proteins (see Section III), but would be initiated conversely by the same toxin rather directly in certain cell types. The lysine residues in Dimer-2 are responsible for this *divalent* type of binding of the B-oligomer. Pertussis toxin was not mitogenic any longer but still elicited ADP-ribosylation of G proteins in intact cells, if the lysine residues in the constituent Dimer-2 had been chemically modified, e.g., acetamidinated[25,26] or dimethylated.[27] Thus, the dimethylated IAP, if available, will be a better probe for G proteins than the native toxin (see References 28 through 31 for review).

TABLE 1
Guanine Nucleotide-Binding α-Subunits of Purified G Proteins

| G Proteins | M_r ($\times 10^{-3}$) of α | | Receptors[a] | Effectors | Localization |
	By SDS-PAGE	Deduced from cDNA			
Substrate for Cholera Toxin					
G_s	45,52	44.5,46	β-Adrenergic	ad-cyclase	—
			Odorant	(activation)	Olfactory cilia
Substrates for Cholera and Pertussis Toxins					
G_T-1	39	40	Rhodopsin	cGMP PDE	Rod
G_T-2	39	40.4	Visual pigments	cGMP PDE(?)	Cone
Substrates for Pertussis Toxin					
G_{i-41}	41	40.4	α-Adrenergic, muscarinic etc.	ad-cyclase (inhibition)	—
G_{i-40}	40	40.5		P-lipase C	
G_{i-43}	43	—		P-lipase A_2(?)	
G_o	39	39		K-channel	Neural cell (?)
G_{HL}	40	—	fMLP (?)	Ca-channel	Neutrophil

Note: Abbreviations used in this table: SDS-PAGE, SDS polyacrylamide gel electrophoresis; ad-cyclase, adenylate cyclase; cGMP PDE, cGMP phosphodiesterase; P-lipase, phospholipase; fMLP, formyl-Met-Leu-Phe.

[a] Only representative receptors are shown.

II. G PROTEINS SERVING AS THE SUBSTRATE OF PERTUSSIS TOXIN-CATALYZED ADP-RIBOSYLATION

A. DIVERSITY OF G PROTEINS

Several G proteins have been purified and proved to be transducers coupled to receptors and/or effectors in membranes or phospholipid vesicles.[32,33] Most of these G proteins are ADP-ribosylated by IAP. G_s involved in adenylate cyclase activation is an exception; it is ADP-ribosylated by cholera toxin instead of pertussis toxin. Transducin (G_T), which mediates light (rhodopsin)-induced activation of cGMP phosphodiesterase in vertebrate retinal cells, is such a unique G protein as to be a substrate for ADP-ribosylation by both pertussis and cholera toxins (Table 1).

These G proteins consist of three polypeptides: a guanine nucleotide-binding α-subunit (M_r = 39,000 to 52,000), a β-subunit (M_r = 35,000 or 36,000), and a γ-subunit (M_r = 5,000 to 10,000). They behave as if they were dimers, however, under physiologic conditions; the βγ-complex is never resolved further to the β- and γ-components unless the protein is denatured by heating in sodium dodecyl sulfate. An association-dissociation cycle between αβγ and α *plus* βγ, which is accompanied by an interconversion between a GDP-bound and a GTP-bound form of the α-component, respectively, affords an important means for the G proteins to play their roles as transducers between receptors and effectors, as will be discussed in detail later (see Section II.B). The βγ-complex is common to most, if not all, of G proteins, each of which hence differs from other members of the G protein family in its α-subunit only. There is a site in the α-subunit to be modified by toxins; an arginine and a cysteine residue are ADP-ribosylated by cholera and pertussis toxin, respectively.

Molecular cloning has revealed the primary structures of nearly all the G proteins that have been purified[34] and has led to a proposal of the tertiary structure of the "average" G protein α-subunit ($G_{av-\alpha}$) based on the crystal structure of a related G protein, the bacterial elongation factor TU.[35] $G_s\alpha$, first defined functionally by its ability to activate adenylate cyclase, was found to be a mixture of two or four peptides with $M_r = 45,000$ and $52,000$.[32] Both arise from a single gene as a result of alternative splicing of internal exons.[34] The real M_r values deduced from sequences of their cDNAs differ somehow from the values previously obtained by SDS polyacrylamide gel electrophoresis (see Table 1).

The α-subunit (α_s) is released from G_s when GTP binds to a GDP-prebound site thereon, and it directly activates the adenylate cyclase catalyst in the membrane. This effect of α_s is not mimicked by the α-subunit of any other G proteins. Cyclic AMP thus generated activates, in turn, a protein kinase in a variety of mammalian tissues and cells. Ca^{2+} mobilization could be thus triggered or more likely modulated via G_s in a type of cells in which the function of the ion channels is regulated by cAMP-dependent phosphorylation of the channel proteins. The involvement of G_s is more direct, however, in the olfactory epithelium. The membrane adenylate cyclase in the olfactory receptor cells is activated by odorants via G_s, and the generated cAMP increases membrane conductance as a result of direct (i.e., without phosphorylation) allosteric modulation of an ion channel.[36] An analogous cyclic nucleotide-gated membrane conductance is found in the rod and cone photoreceptors in which G_T mediates light-induced activation of cGMP phosphodiesterase.[37]

A much more complicated situation is encountered in G_i that mediates inhibition of adenylate cyclase; there are a number of purified G proteins that now fall under this category (Table 1). The G protein first purified as an inhibitor of the cyclase and hence termed G_i has an α-subunit with $M_r = 41,000$, which is ADP-ribosylated by IAP. This G protein is tentatively referred to as G_{i-41} in Table 1. The richest source of this IAP substrate is mammalian brain tissues in which another toxin substrate G protein, G_o, is present in more amounts. The α-subunit of G_o is a peptide with $M_r = 39,000$ on SDS polyacrylamide gels, and the peptide-encoding cDNA has been cloned.[38,39] Additional G proteins ADP-ribosylated by IAP have recently been purified from porcine[40] and bovine[41] brains (first described by Neer et al.[42] and referred to as G_{i-40} in Table 1), human erythrocytes[43] (G_{i-43} in Table 1), and bovine neutrophils[44] as well as HL60 cells that were differentiated into neutrophils[45] (G_{HL} in Table 1). The M_r values are recorded in Table 1 for the α-subunits of these G proteins. In accordance with the diversity of IAP-substrate G proteins, different cDNAs have been obtained by hybridization with oligonucleotide probes based on protein sequences apparently specific to α_i (with $M_r = 41,000$).[38,46-51] G_{i-41}, G_{i-40}, and G_{i-43} in Table 1 must be products of distinct genes. For instance, the cDNAs reported by Nukada et al.[46] and Itoh et al.[38] to correspond to G_α actually code for $G_{i-41}\alpha$ and $G_{i-40}\alpha$, respectively.[52] These $G_i\alpha$ were referred to as the type-1 and type-2α_i, respectively, by Suki et al.[50] and Murphy et al.[53] G_{HL} is probably the same protein as G_{i-40}.[53]

The $\beta\gamma$ component common to all of these pertussis toxin substrates plays an important role in inhibiting adenylate cyclase (see Section II.B). Receptor-mediated inhibitions of the membrane cyclase, or decreases in cellular cAMP, have been reported to be reversed, without exception, by the prior exposure of the membranes or the cells to IAP, the procedure permitting ADP-ribosylation of the mediated G proteins.[54] Thus, G proteins ADP-ribosylated by IAP as listed in Table 1 are candidates for "G_i" that is, in definition, involved in adenylate cyclase inhibition. In fact, both G_{i-41} and G_o were functionally coupled to muscarinic[55-57] and α_2-adrenergic[58] receptors with equal efficiency, when the purified G proteins were reconstituted into phospholipid vesicles together with the purified receptors.

The certain members of this "G_i subfamily" are also candidates for the G proteins that act as transducers between Ca^{2+}-mobilizing receptors and phospholipase C as well as between the other receptors and ion channels, since the receptor-mediated signalings were abolished

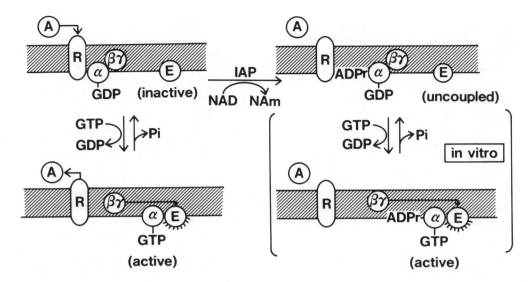

FIGURE 1. Transduction of signal from receptor to effector via G protein, and its blockase by IAP-induced ADP-ribosylation.

by IAP in some cases (see Section III.A and also Reference 59 for review). Furthermore, G_{i-41} was as effective as G_o in stimulating phospholipase C in phospholipid vesicles[60] or in membranes of HL60 cells that had been treated with IAP.[61] Likewise, the G_i purified from erythrocytes opened atrial muscarinic potassium channels[62,63] and termed G_k. G_o was much less effective than the G_i in this regard.

In some cell types stimulated by certain receptor agonists, however, phosphatidylinositol-4,5-bisphosphate was decomposed and inositol-1,4,5-trisphosphate was concurrently generated in the same degree and at the same rate, regardless of whether G proteins in the cells had been ADP-ribosylated or not by IAP[59] Nevertheless, the receptor stimulation increased the rate of GTP hydrolysis, or GTP or its nonhydrolyzable analogs lowered the affinity of agonists for the receptors, in membrane preparations from these cells. These findings are strongly suggestive of an involvement in these cellular responses of a new class of G protein(s) whose function as signal transducer is not impaired by IAP. In this category of G proteins are included p21 (the protein encoded by *ras* oncogenes) and proteins with $M_r = 21,000$ to 32,000 found in membranes of platelets[64,65] placenta,[66] neutrophils,[67] and brains[68] The ADP-ribosylation factor[69] is also a G protein with $M_r = 21,000$. These smaller G proteins are commonly characterized by their insusceptibility to bacterial toxins and by their inability to form a complex with the βγ-component of the bigger G proteins (but see Reference 66). It is not known as yet whether these G proteins in fact act as transducers in signaling systems including those arising from Ca^{2+}-mobilizing receptors.

B. G PROTEINS AS TRANSDUCERS BETWEEN RECEPTORS AND EFFECTORS IN MEMBRANES

Principal properties characteristic and common to all of the G proteins now identified as membrane signal transducers are: (1) an αβγ-trimeric structure, (2) existence of a guanine nucleotide binding site on the α-subunit, (3) exchange of GTP with prebound GDP at the binding site, (4) hydrolysis of GTP to GDP at the same site and (5) ADP-ribosylation of the α-subunit by cholera or pertussis toxin. The trimeric structure of a G protein breaks down and the exchange and hydrolysis of the α-subunit-bound guanine nucleotides are accelerated, as illustrated in Figure 1, upon stimulation by an agonist of the receptor coupled to the G protein.

The upper left of the cartoon in Figure 1 schematizes the membrane receptor situation prior to stimulation of the receptor by an agonist. The receptor with an unoccupied agonist-binding site is complexed with an $\alpha\beta\gamma$-trimeric G protein having GDP at its guanine nucleotide-binding site. The complex formation is a means favoring recognition by the membrane signal proteins of extracellular information, since agonists bind to the complexed receptors with higher affinities than to uncomplexed ones. The G proteins are occasionally considered to be in inactive state under these conditions, since there is no functional interaction between the GDP-bound $\alpha\beta\gamma$-trimer and the effector protein.

Binding of an agonist to the receptor complexed with the trimeric G protein brings about the following sequence of events immediately: (1) dissociation of the $\alpha\beta\gamma$-trimer to an α- and a $\beta\gamma$-subunit; both subunits are then resolved from the receptor, (2) release of GDP from the α-subunit, and (3) binding of GTP to the vacant α-subunit. Combination of (2) with (3) provides a GTP-GDP exchange reaction. A signal is then transmitted to the effector protein via its interaction with the thus formed GTP-bound α- and/or $\beta\gamma$-subunits, as illustrated in the lower left part of Figure 1. Thus, this is what is called the active state of G protein, and the exchange of GTP with prebound GDP on G protein or in membranes has formerly been referred to as the "turnon" reaction. The binding of the next molecule of agonists to receptor is, however, prevented owing to a lowered affinity for the agonist binding under these conditions.

Experimental support for the sequence is as follows, though unpublished data (Katada and Ui) are included. GDP was released from isolated α_{41} ten times as readily as from $\alpha_{41}\beta\gamma$, implying that resolution of the trimeric G protein to the monomeric α-peptide precedes the release of GDP therefrom. The guanine nucleotide-binding site on a trimeric G protein can be emptied of GDP by transient denaturation in concentrated ammonium sulfate followed by renaturation. Binding of GTP to the native GDP-bound G protein was, but the binding to the thus prepared GDP-free G protein was not, accelerated by stimulation by carbachol of muscarinic receptors when the purified receptors were reconstituted into phospholipid vesicles together with these G proteins. Thus, the primary effect of receptor stimulation on the coupled G protein is to release GDP therefrom probably due to a decrease in the affinity of GDP binding upon the receptor-induced resolution of the trimer into the monomer. The vacant site is subsequently occupied by intracellular GTP to be reflected in a well-known receptor-coupled GTP-GDP exchange reaction.[16]

The GTP once bound to the α-subunit of a G protein is gradually hydrolyzed to GDP due to an intrinsic GTPase activity of this peptide; the GDP-bound α reassembles with $\beta\gamma$ to regain the initial inactive state of the G protein complexed with a receptor. Hence, the GTP hydrolysis on G protein serves as a "turnoff" reaction. The receptor-G protein complex is then ready for a next signal delivered by another molecule of agonist. Consecutive repetition of the "turnon" and "turnoff" reactions thus constitutes cycles of signal transduction from receptors to effectors via G proteins in the membrane.

C. UNCOUPLING OF G PROTEINS FROM RECEPTORS AFTER BEING ADP-RIBOSYLATED BY PERTUSSIS TOXIN

The $\alpha\beta\gamma$-trimer of a G protein is the only molecular species that serves as the substrate of IAP-catalyzed ADP-ribosylation. The α-subunits isolated from pertussis toxin substrate G proteins are never ADP-ribosylated by the toxin, regardless of whether the guanine nucleotide-binding sites thereon are occupied by GTP (GTPγS) or GDP, or unoccupied by any one. The cysteine located four residues from the carboxyl terminus of the α-subunit is the site ADP-ribosylated by IAP. No cysteine residue is present at this site in such a G protein as G_s that is not ADP-ribosylated by the toxin.

When this cysteine residue of a G protein was ADP-ribosylated in cell membranes or in phospholipid vesicles into which the G protein and purified receptor had been reconstituted, there were no increases in the activities of the G protein, such as those being reflected in the binding of GTPγS, the release of GDP in exchange for GTP and the hydrolysis of GTP, upon receptor stimulation.[56,70] No signal was transmitted from receptors to effectors including adenylate cyclase, phospholipase C and potassium channels under these conditions. Likewise, the affinity for agonist binding to receptors in membranes or phospholipid vesicles was lowered by ADP-ribosylation of coexisting G_i or G_o, provided GTP analogs were absent in the ambient medium.[14] These experimental results indicate that once G proteins are ADP-ribosylated by IAP they are never capable of being coupled to receptors any longer.[54]

The effect of IAP-induced ADP-ribosylation to uncouple the modified G proteins from receptors was extraordinarily specific. None of the functions of purified G_i or G_o observed *in vitro* were affected by ADP-ribosylation unless these functions were enhanced by stimulation of coexisting receptors. For instance, GTP bound to, it was hydrolyzed to GDP on, and the GDP was then released in exchange for another molecule of GTP from, purified G proteins at essentially the same rates either after or before they were ADP-ribosylated, when these kinetic data were obtained in the absence of coupled receptors. Moreover, purified G_i or G_o ADP-ribosylated by IAP was resolved into an α and a βγ-subunit upon the addition of GTPγS or fluoride ions at the same rate as was modified G_i or G_o.[71] GTPγS-bound ADP-ribosylated $G_i\alpha$ thus obtained competed with $G_s\alpha$ for a site on the catalytic protein of adenylate cyclase at the same potency and efficacy as did the GTPγS-bound unmodified $G_i\alpha$.[72] The competition between $G_{s\alpha}$ and $G_{i\alpha}$ for the cyclase catalyst is one of the mechanisms for G_i-induced inhibition of the cyclase,[72] though calmodulin-mediated inhibition has recently been proposed as the more likely mechanism.[73]

Thus, the cysteine four residues from the carboxyl terminus of the α-subunit of a G protein is indispensable for its coupling to receptors in membranes. Probably, the carboxyl terminal several amino acid residues of the α-subunit forms at least part of the contact with receptor elements coupled to the G protein. The modification of this cysteine residue interferes with coupling of the G protein to receptors very selectively as illustrated in the right hand part of Figure 1. In fact, the selective alkylation of the same cysteine residue by a low concentration of *N*-ethylmaleimide (NEM) exerted just the same influence on the interaction of the G protein with receptors as did IAP-catalyzed ADP-ribosylation.[74,75] Alkylation of G proteins by NEM would be somehow better than enzymic ADP-ribosylation, as a means to uncouple the G proteins from receptors in fragile signaling machinery, since the alkylation proceeds more rapidly than ADP-ribosylation under mild conditions (e.g., in an ice bath).

III. PERTUSSIS TOXIN SUBSTRATE G PROTEINS MEDIATING SIGNALING OTHER THAN THE ADENYLATE CYCLASE SYSTEM

A. PERTUSSIS TOXIN-SUSCEPTIBLE PHOSPHOLIPASE C ACTIVATION

The membrane receptor having seven putative transmembrane segments in its single peptide chain is coupled to a G protein that transduces signals from the receptor to the effector within the membrane. The adenylate cyclase catalysis is the first membrane effector that proved to be controlled directly via G proteins. G_i involved in inhibition of the cyclase catalyst has been identified as the specific target of IAP as described in Section I.B.

A possible involvement of G proteins in signaling systems in which cAMP does not act as the principal intracellular messenger was proposed in 1983[76,77] taking advantage of IAP which should abolish the involvement of G proteins in these non-cAMP systems as well. Prior treatment of mast cells with the toxin abolished an agonist-induced release of histamine from the cells. The abolition was not accounted for by accompanying minor changes in the cellular content of cAMP.[76] α_1-Adrenergic receptor-coupled promotion of phosphatidylinositol turnover was less marked in adipocytes from IAP-treated rats than in the cells from control (nontreated) cells.[77] These observations were supported by the report in the same year that Ca^{2+} added to the extracellular medium triggered histamine secretion from mast cells only if the cells had been rendered permeable to, and actually loaded with, GTP analogs.[78] The target of the GTP analogs injected into the cells must have been one of the IAP substrates, since much less histamine was released under these conditions from the cells that had been exposed to the toxin than from the nonexposed cells.[79]

One of the cell types so far studied extensively as to the susceptibility of their non-cAMP signaling to IAP is neutrophils possessing membrane receptors highly sensitive to the chemotactic peptide, formylmethionyl-leucyl-phenylalanine (fMLP). Various kinds of responses of neutrophils, or HL-60 cells that had been differentiated into neutrophil-type cells, to fMLP[80-91] platelet-activating factor[92,93] and other receptor agonists[93] were markedly diminished or mostly abolished when membrane G proteins were ADP-ribosylated by IAP.[59,94] The site with which the toxin-substrate G protein(s) interacts was located between the receptors and phospholipase C; the toxin treatment interfered with the breakdown of phosphatidylinositol 4,5-bisphosphate, the specific substrate of the enzyme, and the generation of inositol triphosphates, the products of the enzymic reaction, occurring upon stimulation of the receptors in intact cell,[81,85,86,90,92] digitonin-permeabilized cell,[95-97] and cell-free membrane preparations[61,98-101] from neutrophils and platelets. There have been a number of additional reports showing that receptor-mediated inositol phospholipid turnover or intracellular Ca^{2+} mobilization (or cellular responses probably arising therefrom) was blocked by IAP in other cells as well. They include the responses of mast cells to compound 48/80,[102] macrophages to leukotriene B[103], smooth muscle cells to norepinephrine,[104] histamine,[104] vasopressin,[105] serotonin, and angiotensin II,[106] endothelial cells to leukotriences,[107] NG108-15[108] and renal papillary collecting tubule[109] cells to bradykinin, human-mouse hybrid cells to fMLP,[110] renal mesangial cells to angiotensin II,[111] and human melanoma cells to an autocrine factor.[112] A new type of glutamate receptor occurring in *Xenopus* oocytes after injection with rat-brain mRNA activated phospholipase C and mobilized Ca^{2+} through the mediation of an IAP-substrate G protein.[113]

More direct evidence for G protein-induced activation of phospholipase C has recently been afforded by means of reconstitution of purified IAP-substrate G proteins into the toxin-treated HL-60 membranes[61] or into the phospholipid vesicles containing the purified platelet enzyme.[60] Hydrolytic cleavage of cellular G proteins was also suggestive of their roles in phospholipase C activation in platelets.[114] Not only G_o but also G_{i-41} or G_{i-40} purified from brain membranes was found to be capable of being functionally coupled to phospholipase C in the reconstitution systems.[60,61] Purified G_{i-41} was coupled to fMLP receptors in neutrophil membranes as such, but not after being ADP-ribosylated by IAP.[115] Multiple forms of phosphatidylinositol-specific and Ca^{2+}-activatable phospholipase C have been purified from brains, platelets, and other tissues (see Reference 116). A possible association of these enzymes with G proteins has been suggested.[117,118] Thus, it is now widely accepted, on the basis of the reports cited above, that the IAP-substrate G protein couples certain Ca^{2+}-mobilizing receptors to activation of phospholipase C in certain cell types.[59]

D_2 dopaminergic agonists inhibited the angiotensin II-stimulated inositol phosphate generation in primary culture of anterior pituitary cells.[119] No inhibition was observed, however, in the cells in which the α-subunits of G_{i-41}, G_{i-40}, and G_o had been ADP-ribosylated by IAP. This is probably the first paper suggesting that a toxin substrate G-protein could mediate a receptor-linked inhibition, rather than an activation, of phospholipase C in cells.

B. PERTUSSIS TOXIN-SUSCEPTIBLE PHOSPHOLIPASE A_2 ACTIVATION

Ca^{2+} is the major regulator of phospholipase A_2.[120] Ca^{2+} mobilized by the products of the phospholipase C reaction is responsible for the release of archidonic acid that follows the stimulation of Ca^{2+}-mobilizing receptors. The arachidonic acid release should be, therefore, inhibited by IAP in such cell types in which the toxin substrate G-proteins couple receptors to phospholipase C as exemplified in Section III.A. Such was really the case with neutrophils stimulated by fMLP[80] and with mast cells stimulated by compound 48/80.[102]

Arachidonic acid can be produced not only directly by the action of phospholipase A_2 but also by sequential action of phospholipase C and diacylglycerol lipase. Receptor-mediated production of arachidonic acid in thyroid cells,[121] platelets,[122] or mast cells[123] was not inhibited by neomycin, an inhibitor of phospholipase C, or RHC80267, an inhibitor of diacylglycerol lipase, suggesting that the receptor stimulation can lead to rather direct activation of phospholipase A_2. Involvement of an IAP-substrate G-protein in the phospholipase A_2 activation was first proposed with 3T3 fibroblasts in which receptor-linked increases in arachidonic acid release were prevented by the toxin treatment of cells,[124] despite the failure of the same toxin treatment to inhibit inositol release reflecting phospholipase C activation.[125] Conceivably, receptor-coupled activations of phospholipase C and A_2 are mediated by different G proteins including IAP-sensitive and -insensitive ones.[126-128] In bovine rod outer segements, the α-subunit of transducin activates cGMP phosphodiesterase, while its $\beta\gamma$-subunits are involved in phospholipase A_2 activation.[129-130] The activation of phospholipase C or A_2 may produce intracellular signals that are responsible for cell proliferation. A posssible involvement of IAP-substrate G protein in the proliferative signals will be discussed later (see Section III.E).

C. PERTUSSIS TOXIN-SUSCEPTIBLE POTASSIUM CHANNEL ACTIVATION

A variety of cells have been shown to possess Ca^{2+}-dependent K^+ channels that are sequentially activated as the intracellular Ca^{2+} concentration rises progressively. Receptor-linked and G protein-mediated activation of phospholipase C can thus lead to increases in the K^+ channel activity in these cells.[131-133] The decrease in cellular cAMP via an IAP-susceptible G protein was also responsible for an outward K^+ current in neural cells.[134] More direct coupling, i.e., mediated by neither Ca^{2+}, cyclic nucleotides, nor other intracellular messengers, has lately been reported for M_2-muscarinic receptors that open K^+ channels in mammalian atrial cells.[135-139] The K^+ channels were uncoupled from the receptors in membranes in which IAP-substrate G proteins had been ADP-ribosylated by the toxin. G proteins also coupled K^+ channels to adenosine,[140] dopamine,[141] histamine,[141] serotonin,[142] $GABA_B$,[142] glutamate$_B$,[143] somatostatin,[144] and μ- and δ-opiate[145] receptors, although an intramembrane (and non-cAMP) second messenger, such as an arachidonic acid metabolite observed in sensory neurons of *Aplysia*,[146] might be involved in some cases.

One of the G_i proteins that had been purified from human erthrocytes as an IAP substrate opened K^+ channels in atrial cells with a much higher potency than did G_s or G_o.[62] A question as to whether the α-subunit[63,139] or the $\beta\gamma$-component[148] of the toxin-substrate G protein is primarily involved in the opening of K^+ channels is still a subject of controversy.[149,150]

D. PERTUSSIS TOXIN-SUSCEPTIBLE CALCIUM CHANNEL REGULATION.

Voltage-operated Ca^{2+} channels have been classified into T- (transient), N- (neuronal) and L- (long-lasting) channels.[151] Receptor-mediated regulation of these channels can be explained by either a direct interaction of receptor-coupled G proteins with the channel (or with the unidentified channel-associated protein) or an indirect regulation of the channel through the G-protein-dependent generation of second messengers such as cAMP, cGMP, inositol-P_3/inositol-P_4, or diacylglycerol.[152] The L-type channels in cardiac and other tissues were in fact phosphorylated by cAMP-dependent protein kinase (see reference 151 for review). Cyclic AMP and cGMP were effective in gating Ca^{2+}/Na^+ channels allosterically (i.e., without phosphorylation) in olfactory cilia.[153,154]

Stimulation of Ca^{2+} channels via an IAP-substrate G protein, i.e., the inhibition of Ca^{2+} influx by IAP, was observed when the influx was induced by angiotensin II in adrenal cortical cells[152,155] or evoked by depolarization in NG108-15 neuroblastoma-glioma hybrid cells.[156] Phorbol ester-induced Ca^{2+} influx into human neutrophils observable in Na^+-free medium was also largely abolished by prior treatment of the cells with IAP.[157] Mammalian cardiac Ca^{2+} channels, however, were activated via G_s, rather than G_i or other IAP-substrate G proteins, without involvement of protein phosphorylation.[158] An IAP substrate was also involved in stimulatory responses of Ca^{2+} channels to the channel ligands (i.e., Ca^{2+} channel blocking agents) at hyperpolarized membrane potentials.[159]

The larger number of publications have reported G protein-mediated inhibition, rather than the activation, of Ca^{2+} influx through voltage-gated channels. The inhibition by somatostatin,[160-165] carbachol,[161] norepinephrine,[166] GABA,[166] dopamine,[167] acetylcholine,[168] or an opioid of the δ-type[169] was at least in part independent of cellular cAMP but nevertheless was reversed when the membrane G proteins were ADP-ribosylated by IAP in anterior pituitary,[160-165,167] ganglion neurones,[166,168] or in NG108-15 cells.[169] Neuronal activities were inhibited upon stimulation of various receptors again in an IAP-sensitive manner.[170-174] In IAP-treated NG108-15 cells, the inhibitory action of a synthetic opioid was restored by injecting purified IAP-substrate G proteins into the toxin-treated cells.[169] The α-subunit of G_o was ten times more potent than the α-subunit of G_i, while their βγ-subunits or transducin, was without effect, in this regard.

Thus, by analogy with their roles in receptor-mediated regulation of second messenger-producing membrane enzymes, IAP-substrate G-proteins may act as transducers between cell surface receptors and voltage-dependent Ca^{2+} channels, mediating not only hormonal stimulation but also inhibition of the ion channel activity.

E. PERTUSSIS TOXIN-SUSCEPTIBLE CELL PROLIFERATION AND DIFFERENTIATION

Stimulation of growth factor receptors in quiescent cells or cells arrested at the G_0/G_1 phase gives rise to rapid generation of intracellular signals that are responsible for the delayed onset of DNA synthesis or the G_0/G_1-to-S-phase transition. Following the first indication[59] in 1986 that IAP-susceptible G proteins may mediate certain growth factor-induced proliferation of Swiss 3T3 fibroblasts, three papers reported the inhibition by IAP of DNA synthesis or cell growth induced by serum,[175] thrombin,[176] or bombesin[177] in the same or another clone of fibroblasts which are highly responsive to these growth factors. The toxin-induced inhibition of DNA synthesis or of the generation of intracellular signals eventually leading to the initiation of the synthesis in these and other cell types is characterized by several striking features as follows.

First, mitogenic action of some growth factors was inhibited, but the action of other factors was not, by IAP treatment of the same cells under the same conditions, suggesting that there are dual signaling pathways, IAP-sensitive and -insensitive ones, leading to cell proliferation. For instance, the addition of thrombin to G_o-arrested Chinese hamster CCL39 fibroblasts provoked activation of phospholipase C, stimulation of amiloride-sensitive Na^+/H^+ exchange and DNA synthesis in an IAP-sensitive manner, while EGF (epidermal growth factor), FGF (fibroblast growth factor), or insulin increased DNA synthesis in the toxin-insensitive manner.[176,178,179] The phospholipase C activation was the earliest event occurring after the thrombin receptor stimulation, but it did not occur when FGF acted as mitogen in this cell line,[180] although the subsequent signals (i.e., increases in cytoplasmic Ca^{2+} and pH, phosphorylation of S6 protein, and expression of c-*fos* and c-*myc* mRNA) were common following stimulation of both types of receptors. Activation of membrane G proteins by AlF_4^- also caused production of inositol trisphosphates again in an IAP-sensitive manner in this cell line.[178,181] Thus, the results suggest that an IAP-substrate G protein couples the thrombin receptor to phospholipase C that sets in motion an intracellular signaling pathway responsible for cell proliferation.

Similar differential inhibitory effects of IAP were observed with other receptors in other cell types. For instance, bombesin-induced DNA synthesis was, but the EGF-, PDGF-, or insulin-induced synthesis was not, inhibited by IAP in Swiss 3T3 fibroblasts.[177,182,183] Leukotriene-induced synthesis of prostacyclin (a mediator of cell replication[184]) was, but bradykinin-induced synthesis thereof was not, inhibited by the toxin in bovine endothelial CPAC cells.[107] On the contrary, EGF activated phospholipase C and mobilized Ca^{2+} in an IAP-sensitive manner in hepatocytes which responded to angiotensin II in the toxin-insensitive manner.[185-189] Insulin-like growth factor (IGF)-I and -II were potent mitogens and stimulated Ca^{2+} influx in Balb/c 3T3 fibroblasts rendered competent with platelet-derived growth factor (PDGF) and primed with EGF.[190-192] These actions of IGF-I and -II were totally abolished by prior exposure of the cells to IAP.[191,192]

Second, the ADP-ribosylation of cellular G proteins by IAP caused inhibition of only a fraction of the sequential cascade of intracellular signaling events that are considered to trigger the cell growth or the G_1/G_o-to-S-phase transition eventually. Thrombin, bradykinin, angiotensin II, phosphatidic acid, and prostaglandin $F_2\alpha$ were all mitogens in Swiss 3T3 fibroblasts when added together with insulin or EGF.[193] All of these receptor agonists caused decreases in cAMP due to adenylate cyclase inhibition, activation of phospholipase C, increases in arachidonic acid release, Ca^{2+} influx, and activation of ouabain-sensitive Na^+-pump probably following alkalinization of cells by accelerated Na^+/H^+ exchange.[124,125] The exposure of this cell line to IAP caused ADP-ribosylation of G proteins with the molecular mass around 40kDa. Incidentally, the agonist-induced inhibition of adenylate cyclase, phosphatidic acid release, Ca^{2+} influx, and DNA synthesis were abolished or markedly diminished by the toxin treatment without significant attenuation of the phospholipase C or Na^+-pump activations.[124,125,193] The insusceptibility to IAP of receptor-mediated phospholipase C activation in 3T3 cells was also reported elsewhere.[182,183] A simple explanation for the failure of IAP to affect certain signaling events despite inhibition of the eventual DNA synthesis is that an IAP-substrate G protein mediates a step that is distal, rather than proximal to receptor, in the signaling cascade, or that the IAP-insensitive events are not included in the signaling cascade essential for the initiation of DNA synthesis. Thus, application of IAP to the long cascade of signaling events such as those involved in cell proliferation may contribute to our understanding not only as to which events are mediated by IAP substrates but also of the relative importance of the events observed in cellular signaling.

Among the earliest biochemical changes induced by growth factors, a ubiquitous cellular response is an increase in intracellular pH due to the stimulation of an amiloride-sensitive Na^+/H^+ exchanger.[194-196] Although the phosphorylation of the Na^+/H^+ antiporter by protein kinase C was a likely explanation first proposed for the growth factor-induced activation of Na^+/H^+ exchange,[194,195] it is also likely that the stimulation of the antiporter precedes the phospholipase C and A_2 activations (and hence the protein kinase C activation) in signaling cascades arising from α_2-adrenergic receptors in platelets[197-199] and NG108-15 cells.[200] In fact, the α_2-adrenergic receptor purified from porcine brain[201] was tightly associated with the Na^+/H^+ exchanger activity.[202] Stimulation of these types of receptors (α_2-adrenergic, opiate, and muscarinic receptors) in NG108-15 cells caused both adenylate cyclase inhibition and alkalinization of the cell interior due to Na^+/H^+ exchange. Interestingly, the inhibition of adenylate cyclase was, but the acceleration of Na^+/H^+ exchange was not, blocked by IAP,[203] suggesting that the inhibition of adenylate cyclase via an IAP-substrate G protein is not responsible for intracellular alkalinization that appears to play an important role in the activation of phospholipase A_2[203] and the initiation of cell growth.[184] In the case of vascular smooth muscle cells, however, thrombin-induced and protein kinase C-independent acceleration of Na^+/H^+ exchange was abolished by the treatment of the cells with IAP as was the case with CCL39 cells (see above).[204] The Na^+/H^+ antiporter is activated probably via different pathways in different cell types.

Third, the proliferative response of 3T3 fibroblasts to thrombin or other growth factors were attenuated by IAP that was added to the culture medium 1 to 3 h after the addition of the growth factors.[193] An interval of several hours is indispensable between the addition of IAP to the extracellular fluid of any type of cells and the completion of ADP-ribosylation of G proteins within the membrane of the cells (see Section I.C). Thus, an IAP-substrate must be involved in the delayed onset of intracellular signals or the sustained duration of signals arising from growth factor receptor stimulation. It is very likely that the toxin-substrate G protein(s) play an important role in the intracellular signaling in addition to its role as a membrane transducer of receptors for extracellular signal substances including growth factors.

Fourth, multiple roles of G proteins have been suggested, in accordance with the above-mentioned notion, in cells under particular culture conditions. Increased cellular levels of cAMP and stimulation of protein kinase C by diacylglycerol, the product of receptor-coupled phospholipase C, are both mitogenic signals in synergy with other growth factors.[205] Actually, cholera toxin producing vast amounts of cAMP and phorbol ester activating protein kinase C were potent mitogens in Swiss 3T3 fibroblasts, when they were combined with insulin, a progression factor.[193] The DNA synthesis induced by these mitogens was not inhibited by IAP which was effective when the cells were rendered competent by thrombin or other receptor agonists.[193] This appeared to be reasonable if IAP interacted with a G protein communicating between the receptor and effector in membranes; the mediation by the G protein should be bypassed by cholera toxin or phorbol ester. Distinctly different situation was produced, however, by combination of cAMP with protein kinase C activation. The generation of cAMP by forskolin or cholera toxin was markedly enhanced by phorbol ester or receptor agonists involved in diacylglycerol production in the same cell line.[206] The enhancement was mostly abolished by the IAP treatment of the cells, despite the failure of the toxin to affect the protein kinase C activation by phorbol ester of diacylglycerol. Thus, the signaling step with which IAP interacts under these conditions must be distal to protein kinase C.[206] Indeed, an action of Ca^{2+} ionophore, A23187, to accelerate prostaglandin generation was influenced by IAP treatment of bovine aortic endothelial cells.[207]

The involvement of an IAP-substrate G protein in generation of signals related to neither phospholipase C activation, Ca^{2+} mobilization, nor protein kinase C activation was also suggested in the chemotactic peptide-induced respiratory burst in human neutrophils that had been depleted of Ca^{2+} and primed with subthreshold doses of phorbol ester.[208] The peptide activated NADPH oxidase very markedly in these cells without detectable increases in phosphoinostitide turnover and intracellular Ca^{2+} concentration. The activation was totally blocked by IAP,[209] suggesting an involvement of a G protein in a Ca^{2+}-independent signaling pathway. The signaling pathway, distinct from changes in the intracellular Ca^{2+} concentration and activation of protein kinase C, has recently been proposed to occur in electropermeabilized neutrophils as well.[210] Similarly, thrombin-induced histamine secretion from bone marrow-derived mouse mast cells was abolished when the cells were treated with IAP, despite the fact that the simultaneously occurring phosphoinositide turnover, Ca^{2+} mobilization or arachidonic acid release was not affected by the toxin treatment.[211] A high concentration of thrombin gave rise to serotonin secretion from platelets in an IAP-sensitive manner, even if Ca^{2+} mobilization was inhibited by the injection of a GDP analog into the permeabilized cells.[97] In accordance with these observations, suggestions have been made for rather direct involvement of G proteins in exocytotic mechanism, although the proteins suggested to be involved did not appear to serve as the substrate of IAP-catalyzed ADP-ribosylation.[212-214]

In conclusion, a long and complicated cascade of intracellular signals appears to be required for cells to proliferate or differentiate in response to membrane receptor stimulation. In addition to their well-known function as transducers between certain receptors and certain effectors in membranes, IAP-substrate G proteins may mediate distal step(s) of intracellular signaling pathways, although several lines of evidence so far provided are only indirect, preliminary or circumstantial. Vitamin A-induced differentiation of guinea pig macrophages was also prevented by the exposure of cells to IAP.[215] A probable involvement of G proteins in the gene expression will be an additional important subject of further research.

IV. CONCLUSION

Most of the extracellular signals are recognized by plasma membrane receptor systems in mammalian cells. There are a lot of the membrane receptor systems each of which is composed of three protein components; i.e., a receptor, a transducer and an effector. G proteins with a typical $\alpha\beta\gamma$-trimeric structure are the only transducers so far identified definitely. G protein-coupled effectors include adenylate cyclase (activated or inhibited by the coupled G protein), cGMP phosphodiesterase (activated), phospholipase C (mostly activated), potassium channel (activated), calcium channel (activated or inhibited) and probably phospholipase A_2 (activated).

The universally adopted criterion for receptor-G protein coupling is the effect of GTP (or GTP analogs) to decrease the affinity for agonist binding to the receptor. The rationale of this criterion was presented in Section II.B in this chapter as illustrated in Figure 1. This "GTP-effect" was confirmed for essentially all of the receptors regulating the above-mentioned effectors. Pertussis toxin, IAP, affords an additional and more reliable criterion. The blockade by IAP of a receptor-mediated signaling provides evidence for the involvement of an IAP-substrate G proteins in the signaling, since the G protein is not capable of being coupled to any receptor, once it has been ADP-ribosylated by the toxin (Figure 1).

The IAP-substrate G protein first discovered was that involved in the inhibition of adenylate cyclase. Thus far, receptor-mediated inhibitions of adenylate cyclase or decreases of cellular cAMP due to the cyclase inhibition, have been reported, essentially without exception, to be abolished if the involved G proteins were ADP-ribosylated. The IAP-insensitive decreases in cellular cAMP, if observed so far, appeared to occur by mechanisms other than G protein-dependent adenylate cyclase inhibition.[216,217] The important problem then to be solved in the near future will be which is the real transducer among four kinds of IAP-substrate G proteins so far identified, G_i-1 (G_{i-41}), G_i-2 (G_{i-40}), G_i-3, and G_o. The transducers regulating K^+ and Ca^{2+} channels and probably phospholipase A_2 also appear to be IAP-substrate G proteins in most cases. Thus, IAP will be still a good probe for the receptors regulating these IAP-susceptible effectors.

The situation is currently complicated, however, for G proteins involved in phospholipase C activation. Apart from a number of publications reporting the IAP-induced blockade of receptor-mediated activation of phospholipase C (see Section III.A), there also are reports that are unfavorable for the IAP susceptibility of the same enzyme activation (see reference 59 for review); IAP was without effect on phospholipase C activation in certain cases. It is suggested, therefore, that G proteins(s) other than IAP substrates are involved in phospholipase C activation in these cases. The candidates of such non-IAP-substrate transducers are p21, the product of *ras*-oncogene, and the GTP-binding proteins with similar small molecular weights recently purified from brains,[218,219] since these proteins are not ADP-ribosylated by IAP. No evidence is available, however, that these smaller GTP-binding proteins form a complex with a receptor protein with a high affinity for agonist binding, despite the fact that the above-mentioned "GTP-effect" has been observed even for the receptors coupled to phospholipase C in an IAP-insensitive manner.

Thus, an alternative possibility cannot be excluded that IAP-substrate G proteins also mediate IAP-insensitive activation of phospholipase C. Not all of the IAP-substrates involved might be ADP-ribosylated in intact cells under the usual conditions. Stimulation of muscarinic receptors by carbachol caused both inhibition of adenylate cyclase and activation of phospholipase C in embryonic chicken heart cells.[220] The potency of carbachol was 100-fold greater for inhibiting adenylate cyclase than for stimulating phosphatidylinositol breakdown in this cell type. The dose-response curves for agonist binding were two log units to the right of those for agonist inhibition of adenylate cyclase and close to those for agonist activation of phospholipase C. Thus, only a fraction of muscarinic receptors has to be occupied to cause inhibition of adenylate cyclase leaving the residual ones as "spare" receptors, while essentially all the fractions of receptor are required for activation of phospholipase C. Similar different potencies of the muscarinic agonist between adenylate cyclase inhibition and phospholipase C activation were reproduced in Chinese hamster ovary (CHO) cells in which a single type of recombinant M_2 muscarinic receptors were expressed.[221] The much larger number of receptors had to be stimulated by agonist for phospholipase C to be fully activated than for adenylate cyclase to be totally inhibited. In fact, 100-times as high a concentration of IAP was required for the total blockade of phospholipase C activation as for the abolition of adenylate cyclase inhibition in CHO cells having recombinant M_2-receptors. It would be likely that only a small compartment of G protein pools in membranes that has survived the IAP-catalyzed ADP-ribosylation procedure could transduce a fraction of dense signals to phospholipase C, but would not be involved in adenylate cyclase inhibition that is achieved via only a small fraction of the receptor pool.

In any event, islet-activating protein (IAP), pertussis toxin, will continue to make greater and greater contributions to future studies on the precise role of G proteins in a variety of cellular signaling pathways.

ACKNOWLEDGMENTS

A part of the content of this review article depends on the work supported by research grants from the Scientific Research Fund of the Ministry of Education, Science and Culture, Japan, and by research grants from Yamada Scientific Foundation and from Toray Science and Technology Grants in Japan.

REFERENCES

1. **Sumi, T. and Ui, M.**, Potentiation of the adrenergic *beta*-receptor-mediated insulin secretion in pertussis-sensitized rats, *Endocrinology*, 97, 352, 1975.
2. **Katada. T. and Ui, M.**, Accelerated turnover of blood glucose in pertussis-sensitized rats due to combined actions of endogenous insulin and adrenergic β-stimulation,, *Biochim, Biophys. Acta*, 421, 57, 1976.
3. **Katada. T. and Ui, M.**, Spontaneous recovery from streptozotocin-induced diabetes in rats pretreated with pertussis vaccine or hydrocortisone, *Diabetologia*, 13, 521, 1977.
4. **Katada, T. and Ui, M.**, Perfusion of the pancreas isolated from pertussis-sensitized rats: potentiation of insulin secretory responses due to β-adrenergic stimulation, *Endocrinology*, 101, 1247, 1977.
5. **Toyota, T., Kakizaki, M., Yajima, M., Okamoto, T., and Ui, M.**, Islet-activating protein derived from the culture supernatant fluid of *Bordetella pertussis:* effect on spontaneous diabetic rats, *Diabetologia*, 14, 319, 1978.
6. **Yajima, M., Hosoda, K., Kanbayashi, Y., Nakamura, T., Nogimori, K., Nakase, Y., and Ui, M.**, Islet-activating protein in *Bordetella pertussis* that potentiates insulin secretory responses of rats. Purification and characterization, *J. Biochem.*, 83, 295, 1978.
7. **Yajima, M., Hosoda, K., Kanbayashi, Y., Nakamura, T., Takahashi, I., and Ui, M.**, Biological properties of islet-activating protein purified from the culture medium of *Bordetella pertussis*, *J. Biochem.*, 83, 305, 1978.
8. **Katada, T. and Ui, M.**, Islet-activating protein. Enhanced insulin secretion and cyclic AMP accumulation in pancreatic islet due to activation of native calcium ionophores, *J. Biol. Chem.*, 254, 469, 1979.

9. **Katada, T. and Ui, M.,** Effect of *in vivo* pretreatment of rats with a new protein purified from *Bordetella pertussis* on *in vitro* secretion of insulin: role of calcium, *Endocrinology,* 104, 1822, 1979.

10. **Katada, T. and Ui, M.,** Slow interaction of islet-activating protein with pancreatic islets during primary culture to cause reversal of α-adrenergic inhibition of insulin secretion, *J. Biol. Chem.,* 255, 9580, 1980.

11. **Katada, T. and Ui, M.,** *In vitro* effects of islet-activating protein on cultured rat pancreatic islets. Enhancement of insulin secretion, cyclic AMP accumulation and ^{45}Ca flux, *J. Biochem.,* 89, 979, 1981.

12. **Katada, T. and Ui, M.,** Islet-activating protein. A modifier of receptor-mediated regulation of rat islet adenylate cyclase, *J. Biol. Chem.,* 256, 8310, 1981.

13. **Hazeki, O. and Ui, M.,** Modification by islet-activating protein of receptor-mediated regulation of cyclic AMP accumulation in isolated rat heart cells, *J. Biol. Chem.,* 256, 2856, 1981.

14. **Kurose, H., Katada, T., Amano, T., and Ui, M.,** Specific uncoupling by islet-activating protein, pertussis toxin, of negative signal transduction via α-adrenergic, cholinergic, and opiate receptors in neuroblastoma × glioma hybrid cells, *J. Biol. Chem.,* 258, 4870, 1981.

15. **Murayama, T. and Ui, M.,** Loss of the inhibitory function of the guanine nucleotide regulatory component of adenylate cyclase due to its ADP-ribosylation by islet-activating protein, pertussis toxin, in adipocyte membrane, *J. Biol. Chem.,* 258, 3319, 1983.

16. **Murayama, T. and Ui, M.,** [^{3}H] GDP release from rat and hamster adipocyte membranes independently linked to receptors involved in activation or inhibition of adenylate cyclase. Differential susceptibility to two bacterial toxins, *J. Biol. Chem.,* 259, 761, 1984.

17. **Murayama, T., Katada, T., and Ui, M.,** Guanine nucleotide activation and inhibition of adenylate cyclase as modified by islet-activating protein, pertussis toxin, in mouse 3T3 fibroblasts, *Arch. Biochem. Biophys.,* 221, 381, 1983.

18. **Katada, T. and Ui, M.,** Direct modification of the membrane adenylate cyclase system by islet-activating protein due to ADP-ribosylation of a membrane protein, *Proc. Natl. Acad. Sci. U.S.A.,* 79, 3129, 1982.

19. **Katada, T. and Ui, M.,** ADP-ribosylation of the specific membrane protein of C6 cells by islet-activating protein associated with modification of adenylate cyclase activity. *J. Biol. Chem.,* 257, 7210, 1982.

20. **Rodbell, M.,** The role of hormone receptors and GTP-regulatory proteins in membrane transduction, *Nature,* 284, 17, 1980.

21. **Tamura, M., Nogimori, K., Murai, M., Yajima, M., Ito, K., Katada, T., Ui, M., and Ishii, S.,** Subunit structure of islet-activating protein, pertussis toxin, in conformity with the A-B model, *Biochemistry,* 21, 5516, 1982.

22. **Katada, T., Tamura, M., and Ui, M.,** The A-protomer of islet-activating protein, pertussis toxin, as an active peptide catalyzing ADP-ribosylation of a membrane protein, *Arch. Biochem. Biophys.,* 224, 290, 1983.

23. **Tamura, M., Nogimori, K., Yajima, M., Ase, K., and Ui, M.,** A role of the B-oligomer moiety of islet-activating protein, pertussis toxin, in development of the biological effects on intact cells, *J. Biol. Chem,* 258, 6756, 1983.

24. **Katada, T., Amano, T., and Ui, M.,** Modulation by islet-activating protein of adenylate cyclase activity in C6 glioma cells, *J. Biol. Chem.,* 257, 3739, 1982.

25. **Nogimori, K., Ito, K., Tamura, M., Satoh, S., Ishii, S., and Ui, M.,** Chemical modification of islet-activating protein, pertussis toxin. Essential role of free amino groups in its lymphocytosis-promoting activity, *Biochim. Biophys. Acta,* 801, 220, 1984.

26. **Nogimori, K., Tamura, M., Yajima, M., Ito, K., Nakamura, T., Kajikawa, N., Maruyama, Y., and Ui, M.,** Dual mechanisms involved in development of diverse biological activities of islet-activating protein, pertussis toxin, as revealed by chemical modification of lysine residues in the toxin molecule, *Biochim. Biophys. Acta,* 801, 232, 1984.

27. **Nogimori, K., Tamura, M., Yajima, M., Hashimura, N., Ishii, S., and Ui, M.,** Structure-function relationship of islet-activating protein, pertussis toxin: biological activities of hybrid toxins reconstituted from native and methylated subunits, *Biochemistry,* 25, 1355, 1986.

28. **Ui, M., Nogimori, K., and Tamura, M.,** Islet-activating protein, pertussis toxin: Subunit structure and mechanism for its multiple biological actions, in *Pertussis Toxin,* Sekura, R. D., Moss, J., and Vaughan, M., Eds., Academic Press, New York, 1985, 19.

29. **Nogimori, K., Tamura, M., Nakamura, T., Yajima, M., Ito, K., and Ui, M.,** Dual mechanisms involved in development of diverse biological activities of islet-activating protein, pertussis toxin, as revealed by chemical modification of the toxin molecule, in *Developments in Biological Standardization,* Karger, S., Ed. Dept of Health, Education and Welfare Publication, Rockville, Md, 1985, 51.

30. **Ui, M., Nogimori, K., and Tamura, M.,** Structure of islet-activating protein, pertussis toxin, required for its multiple biological activities, in *Proceedings of 2nd European Workshop on Bacterial Protein Toxins,* Falmagne, P. et al., Eds., Academic Press, New York, 1986, 9.

31. **Ui, M.,** The multiple biological activities of pertussis toxin, in *Pathogenesis and Immunity in Pertussis,* Wardlaw, A. C. and Parton, R., Eds., John Wiley & Sons, London, 1988, 121.

32. **Gilman, A. G.,** G proteins and dual control of adenylate cyclase, *Cell,* 36, 577, 1984.

33. **Stryer, L. and Bourne, H. R.,** G proteins: a family of signal transducers, *Annu. Rev. Cell. Biol.,* 2, 391, 1986

34. **Gilman A. G.,** G Proteins: transducers of receptor-generated signals, *Annu. Rev. Biochem.,* 56, 615, 1987.

35. **Masters, S. B., Stroud, R. M., and Bourne H. R.,** Family of G protein α chains: amphipathic analysis and predicted structure of functional domains, *Protein Engineering,* 1, 47, 1986.

36. **Gold, G. H. and Nakamura, T.,** Cyclic nucleotide-gated conductances: a new class of ion channels mediates visual and olfactory transduction, *Trend Pharmacol. Sci.,* 8, 312, 1987.

37. **Streyer, L.,** Cyclic GMP cascade of vision, *Annu. Rev. Neurosci.,* 9, 87, 1986.

38. **Itoh, H., Kozasa, T., Nagata, S., Nakamura, S., Katada, T., Ui, M., Iwai, S., Ohtsuka, E., Kawasaki, H., Suzuki, K., and Kaziro, Y.,** Molecular cloning and sequence determination of cDNAs coding for α-subunits of G_{s1}, G_{i1}, and G_o proteins from rat brain, *Proc. Natl. Acad. Sci. U.S.A.,* 83, 3776, 1986.

39. **Van Meurs. K. P., Angus, C. W., Lavu, S., Kung, H. -F., Czarnecki, S. K., Moss, J., and Vaughan, M.,** Deduced amino acid sequence of bovine retinal G_o α: similarities to other guanine nucleotide-binding proteins, *Proc. Natl. Acad. Sci. U.S.A.,* 84, 3107, 1987

40. **Katada, T., Oinuma, M., Kusakabe, K., and Ui, M.,** A new GTP-binding protein in brain tissues serving as the specific substrate of islet-activating protein, pertussis toxin, *FEBS Lett.,* 213, 353, 1987.

41. **Mumby, S., Pang, I. -H., Gilman, A. G., and Sternweis, P. C.,** Chromatographic resolution and immunological identification of the $α_{40}$ and $α_{41}$ subunits of guanine nucleotide-binding regulatory proteins from bovine brain, *J. Biol. Chem.,* 263, 2020, 1988.

42. **Neer, E. J., Lok, J. M., and Wolf, L. G.,** Purfication and properties of the inhibitory guanine nucleotide regulatory unit of brain adenylate cyclase, *J. Biol. Chem.,* 259, 14222, 1984.,

43. **Iyengar, R., Rich, K. A., Herberg, J. T., Grenet, D., Mumby, S., and Codina, J.,** Identification of a new GTP-binding protein. A M_r = 43,000 substrate for pertussis toxin, *J. Biol. Chem.,* 262, 9239, 1987.

44. **Gierschik, P., Sidiropoulous, D., Spiegel, A., and Jakobs, K. H.,** Purification and immunochemical characterization of the major pertussis-toxin-sensitive guanine-nucleotide-binding protein of bovine neutrophil membranes, *Eur. J. Biochem.,* 165, 185, 1987.

45. **Oinuma, M., Katada, T., and Ui, M.,** A new GTP-binding protein in differentiated human leukemic (HL-60) cells serving as the specific substrate of islet-activating protein, pertussis toxin, *J. Biol. Chem.,* 262, 8347, 1987.

46. **Nukada, T., Tanabe, T., Takahashi, H., Noda, M., Haga, K., Haga, T., Ichiyama, A., Kangawa, K., Hiranaga, M., Mastuo, H., and Numa, S.,** Primary structure of bovine adenylate cyclase-inhibiting G-protein deduced from the cDNA sequence, *FEBS Lett.,* 197, 305, 1986.

47. **Sullivan, K. A., Liao, Y. -C., Alborzi, A., Beiderman, B., Chang, F. -H., Masters, S. B., Levinson, A. D., and Bourne, H. R.,** Inhibitory and stimulatory G proteins of adenylate cyclase: cDNA and amino acid sequences of the α chains. *Proc. Natl. Acad. Sci. U.S.A.,* 83, 6687, 1986.

48. **Michel, T., Winslow, J. W., Smith, J. A., Seidman, J. G., and Neer, E. J.,** Molecular cloning and characterization of cDNA encoding the GTP-binding protein $α_i$ and identification of a related protein, $α_h$ *Proc. Natl. Acad. Sci. U.S.A.,* 83, 7663, 1986.

49. **Didsbury, J. R. and Snyderman, R.,** Molecular cloning of a new human G protein. Evidence for two G_i α-like protein families, *FEBS Lett.,* 219, 259, 1987.

50. **Suki, W. N., Abramowitz, J., Mattera, R., Codina, J., and Birnbaumer, L.,** The human genome encodes at least three non-allellic G proteins with $α_i$-type subunits, *FEBS Lett.,* 220, 187, 1987.

51. **Bray, P., Carter, A., Guo, V., Puckett, C., Kamholz, J., Spiegel, A., and Nirenberg, M.,** Human cDNA clones for an α subunit of G_i signal-transduction protein, *Proc. Natl. Acad. Sci. U.S.A.,* 84, 5115, 1987.

52. **Itoh, H., Katada, T., Ui, M., Kawasaki, H., Suzuki, K., and Kaziro, Y.,** Identification of three pertussis toxin substrates (41, 40 and 39 kDa proteins) in mammalian brain, *FEBS Lett.,* 230, 85, 1988.

53. **Murphy, P. M., Eide, B., Goldsmith, P., Brann, M., Gierschik, P., Spiegel, A., and Malech, H. L.,** Detection of multiple forms of G_i α in HL60 cells, *FEBS Lett.,* 221, 81, 1987.

54. **Ui, M.,** Islet-activating protein, pertussis toxin: a probe for functions of the inhibitory guanine nucleotide regulatory component of adenylate cyclase, *Trends Pharmacol. Sci.,* 5, 277, 1984.

55. **Florio, V. A. and Sternweis P. C.,** Reconstitution of resolved muscarinic cholinergic receptors with purified GTP-binding proteins, *J. Biol. Chem.,* 260, 3477, 1985.

56. **Kurose, H., Katada, T., Haga, T., Haga, K., Ichiyama, A., and Ui, M.,** Functional interaction of purified muscarinic receptors with purified inhibitory guanine nucleotide regulatory proteins reconstituted in phospholipid vesicles, *J. Biol. Chem.,* 261, 6423, 1986.

57. **Haga, K., Haga, T., and Ichiyama, A.,** Reconstitution of the muscarinic acetylcholine receptor. Guanine nucleotide-sensitive high affinity binding of agonists to purified muscarinic receptors reconstituted with GTP-binding proteins (G_i and G_o), *J. Biol. Chem.,* 261, 10133, 1986.

58. **Cerione, R. A., Regan, J. W., Nakata, H., Codina, J., Benovic, J. L., Gierschik, P., Somers, R. L., Spiegel, A. M., Birnbaumer, L., Lefkowitz, R. J., and Caron, M. G.,** Functional reconstitution of $α_2$ adrenergic receptor with guanine nucleotide regulatory proteins in phospholipid vesicles, *J. Biol. Chem.,* 261, 3901, 1986.

59. **Ui, M.,** Pertussis toxin as a probe of receptor coupling to inositol lipid metabolism, in *Receptor Biochemistry and Methodology: Receptor and Phosphoinositides,* Putney, J. W., Ed., Alan R. Liss, New York, 1986, 163.

60. **Banno, Y., Nagao, S., Katada, T., Nagata, K., Ui, M., and Nozawa, Y.,** Stimulation by GTP-binding proteins (G_i, G_o) of partially purified phospholipase C activity from human platelet membranes, *Biochem. Biophys. Res. Commun.,* 146, 861, 1987.

61. **Kikuchi, A., Kozawa, O., Kaibuchi, K., Katada, T., Ui, M., and Takai, Y.,** Direct evidence for involvement of a guanine nucleotide-binding protein in chemotactic peptide-stimulated formation of inositol bis- and tris-phosphate in differentiated human leukemic (HL60) cells: reconstitution with G_i or G_o of the plasma membranes ADP-ribosylated by pertussis toxin, *J. Biol. Chem.,* 261, 11558, 1986.

62. **Yatani, A., Codina, J., Brown, A. M., and Birnbaumer, L.,** Direct activation of mammalian atrial muscarinic potassium channels by GTP regulatory protein G_k, *Science,* 235, 207, 1987.

63. **Codina, J., Yatani, A., Grenet, D., Brown, A. M., and Birnbaumer, L.,** The α subunit of the GTP binding protein G_k opens atrial potassium channels, *Science,* 236, 442, 1987.

64. **Lapetina, E. G. and Reep, B. R.,** Specific binding of [α -^{32}P] GTP to cytosolic and membrane-bound proteins of human platelets correlates with the activation of phospholipase C, *Proc. Natl. Acad. Sci. U.S.A.,* 84, 2261, 1987.

65. **Bhullar, R. P. and Haslam, R. J.,** Detection of 23—27 kDa GTP-binding proteins in platelets and other cells, *Biochem. J.,* 245, 617, 1987.

66. **Evans, T., Brown, M. L., Fraser. E, D., and Northup, J. K.,** Purification of the major GTP-binding proteins from human placental membranes, *J. Biol. Chem.,* 261, 7052, 1986.

67. **Bokoch, G. M. and Parkos, C. A.,** Identification of novel GTP-binding proteins in the human neutrophil, *FEBS Lett.,* 227, 66, 1988.

68. **Hatta, S., Marcus, M. M., and Rasenick M. M.,** Exchange of guanine nucleotide between GTP-binding proteins that regulate neuronal adenylate cyclase, *Proc. Natl. Acad. Sci. U.S.A.,* 83, 5439, 1986.

69. **Kahn, R. A. and Gilman, A. G.,** The protein cofactor necessary for ADP-ribosylation of G_s by cholera toxin is itself a GTP binding protein, *J. Biol. Chem.,* 261, 7906, 1986.

70. **Haga, K., Haga, T., Ichiyama, A., Katada, T., Kurose, H., and Ui, M.,** Functional reconstitution of purified muscarinic receptors and inhibitory guanine nucleotide regulatory protein, *Nature (London),* 316, 731, 1985.

71. **Katada, T., Oinuma, M., and Ui, M.,** Two guanine nucleotide-binding proteins in rat brain serving as the specific substrate of islet-activating protein, pertussis toxin. Interaction of the α-subunits with β γ-subunits in development of their biological activities, *J. Biol. Chem.,* 261, 8182, 1986.

72. **Katada, T., Oinuma, M., and Ui, M.,** Mechanisms for inhibition of the catalytic activity of adenylate cyclase by the guanine nucleotide-binding proteins serving as the substrate of islet-activating protein, pertussis toxin, *J. Biol. Chem.,* 261, 8182, 1986.

73. **Katada, T., Kusakabe, K., Oinuma, M., and Ui, M.,** A novel mechanism for the inhibition of adenylate cyclase via inhibitory GTP-binding proteins. Calmodulin-dependent inhibition of the cyclase catalyst by the $\beta\gamma$-subunits of GTP-binding proteins, *J. Biol. Chem.,* 262, 11897, 1987.

74. **Katada, T., Kurose, H., Oinuma, M., Hoshino, S., Shinoda, M., Amanuma, S., and Ui, M.,** Role of GTP-binding proteins in coupling of receptors and adenylate cyclase, in *Gunma Symposia on Endocrinology,* Vol. 23, VNU Science Press BV, Utrecht, 1986, 45.

75. **Ui, M. and Katada, T.,** Differential roles of G_i and G_o in multiple receptor coupling in brain, in *Current Communications in Molecular Biology,* Michell, R. H. and Putney, J. W., Eds., Cold Spring Harbor Laboratory, 1987, 59.

76. **Nakamura, T. and Ui, M.,** Suppression of passive cutaneous anaphylaxis by pertussis toxin, islet-activating protein, as a result of inhibition of histamine release from mast cells, *Biochem. Pharmacol.,* 32, 3435, 1983.

77. **Moreno, F. J., Mills, I., Gracia-Sainz, J. A., and Fain, J. N.,** Effects of pertussis toxin treatment on the metabolism of rat adipocytes, *J. Biol. Chem.,* 258, 10938, 1983.

78. **Gomperts, B. D.,** Involvement of guanine nucleotide-binding protein in the gating of Ca^{2+} by receptors, *Nature,* 306, 64, 1983.

79. **Nakamura, T. and Ui, M.,** Islet-activating protein, pertussis toxin, inhibits Ca^{2+}-induced and guanine nucleotide-dependent releases of histamine and arachidonic acid from rat mast cells, *FEBS Lett.,* 173, 414, 1984.

80. **Okajima, F. and Ui, M.,** ADP-ribosylation of the specific membrane protein by islet-activating protein, pertussis toxin, associated with inhibition of a chemotactic peptide-induced arachidonate release in neutrophils. A possible role of the toxin substrate in Ca^{2+}-mobilizing biosignaling, *J. Biol. Chem.,* 259, 13863, 1984.

81. **Ohta, H., Okajima, F., and Ui, M.,** Inhibition by islet-activating protein of a chemotactic peptide-induced early breakdown of inositol phospholipids and Ca^{2+} mobilization in guinea pig neutrophils, *J. Biol. Chem.,* 260, 15771, 1985.

82. **Bokoch, G. M. and Gilman, A. G.,** Inhibition of receptor-mediated release of arachidonic acid by pertussis toxin, *Cell,* 39, 301, 1984.

83. **Molski, T. F. P., Naccache, P. H., Marsh, M. L., Kermode, J., Becker E. L., and Sha'afi, R. I.,** Pertussis toxin inhibits the rise in the intracellular concentration of free calcium that is induced by chemotactic factors in rabbit neutrophils: possible role of the "G proteins" in calcium mobilization, *Biochem. Biophys. Res. Commun.,* 124, 644, 1984.

84. **Shefcyk, J., Yassin, R., Volpi, M., Molski, T. E. P., Naccache, P. H., Munoz, J. J., Becker, E. L., Feinstein, M. B., and Sha'afi, R. I.,** Pertussis but not cholera toxin inhibits the stimulated increase in actin association with the cytoskeleton in rabbit neutrophils: role of the "G Proteins" in stimulus-response coupling, *Biochem. Biophys. Res. Commun.,* 126, 1174, 1985.

85. **Brandt, S. J., Dougherty, R. W., Lapetina, E. G., and Niedel, J. E.,** Pertussis toxin inhibits chemotactic peptide-stimulated generation of inositol phosphates and lysosomal enzyme secretion in human leukemic (HL-60) cells, *Proc. Natl. Acad. Sci. U.S.A.,* 82, 3277, 1985.

86. **Bradford, P. G. and Rubin, R. P.,** Pertussis toxin inhibits chemotactic factor-induced phospholipase C stimulation and lysosomal enzyme secretion in rabbit neutrophils, *FEBS Lett.,* 183, 317, 1985.

87. **Lad, P. M., Olson, C. V., and Smiley, P. A.,** Association of the N-formyl-Met-Leu-Phe receptor in human neutrophils with a GTP-binding protein sensitive to pertussis toxin, *Proc. Natl. Acad. Sci. U.S.A.,* 82, 869, 1985.

88. **Satoh, M., Nanri, H., Takeshige, K., and Minakami, S.,** Pertussis toxin inhibits intracellular pH changes in human neutrophils stimulated by N- formyl-methionyl-leucyl-phenylalanine, *Biochem. Biophys. Res. Commun.,* 131, 64, 1985.

89. **Volpi, M., Naccache, P. H., Molski, T. F. P., Shefcyk, J., Huang, C. -K., Marsh, M. L., Munoz, J., Becker, E. L., and Sha'afi, R. I.,** Pertussis toxin inhibits fMet-Leu-Phe- but not phorbol ester-stimulated changes in rabbit neutrophils: role of G proteins in excitation response coupling, *Proc. Natl. Acad. Sci. U.S.A.,* 82, 2708, 1985.

90. **Verghese, M. W., Smith, C. D., and Snyderman, R.,** Potential role for a guanine nucleotide regulatory protein in chemoattractant receptor mediated polyphosphoinositide metabolism, Ca^{++} mobilization and cellular responses by leukocytes, *Biochem. Biophys. Res. Commun.,* 127, 450, 1985.

91. **Lad, P. M., Olson, C. V., and Grewal, I. S.,** A step sensitive to pertussis toxin and phorbol ester in human neutrophils regulates chemotaxis and capping but not phagocytosis, *FEBS Lett.,* 200, 91, 1986.

92. **Naccache, P. H., Molski, M. M., Volpi M., Becker, E. L., and Sha'afi, R. I.,** Unique inhibitory profile of platelet activating factor induced calcium mobilization, polyphosphoinositide turnover and granule enzyme secretion in rabbit neutrophils towards pertussis toxin and phorbol ester, *Biochem, Biophys. Res. Commun.,* 130, 677, 1985.

93. **Lad, P. M., Olson, C. V., Grewal, I. S., and Scott, S. J.,** A pertussis toxin-sensitive GTP-binding protein in the human neutrophil regulates multiple receptors, calcium mobilization, and lectin-induced capping. *Proc. Natl. Acad. Sci. U.S.A.,* 82, 8643, 1985.

94. **Ui, M., Okajima, F., Murayama, T., Nakamura, T., Kurose, H., Itoh, H., and Ohta, H.,** A role of the inhibitory guanine nucleotide-binding regulatory protein in signal transduction via Ca^{2+}-mobilizing receptors, in *Adrenergic Receptors: Molecular Properties and Therapeutic Implications.* Lefkowitz, R. J. and Lindenlaub, E. Eds., F. K. Schattauer Verlag, Stuttgart, 1985, 209.

95. **Lapetina, E. G.,** Effect of pertussis toxin on the phosphodieteratic cleavage of the polyphosphoinositides by guanosine 5'-0-thiotriphosphate and thrombin in permeabilized human platelets, *Biochim, Biophys, Acta,* 884, 219, 1986.

96. **Sklar, L. A., Bokoch, G. M., Button, D., and Smolen, J. E.,** Regulation of ligand-receptor dynamics by guanine nucleotides, *J. Biol. Chem.,* 262, 135, 1987.

97. **Brass, L. F., Laposata, M., Banga, H. S., and Rittenhouse, S. E.,** Regulation of the phosphoinositide hydrolysis pathway in thrombin-stimulated platelets by a pertussis toxin-sensitive guanine nucleotide-binding protein. Evaluation of its contribution to platelet activation and comparisons with the adenylate cyclase inhibitory protein, G_i, *J. Biol. Chem.,* 261, 16838, 1986.

98. **Smith, C. D., Lane, B. C., Kusaka, I., Verghese, M. W., and Snyderman, R.,** Chemoattractant receptor-induced hydrolysis of phosphatidyl-inositol 4,5-bisphosphate in human polymorphonuclear leukocyte membranes. Requirement for a guanine nucleotide regulatory protein, *J. Biol. Chem.,* 260, 5875, 1985.

99. **Anthes, J. C., Billah, M. M., Cali, A., Egan, R. W., and Siegel, M. I.,** Chemotactic peptide, calcium and guanine nucleotide regulation of phospholipase C activity in membranes from DMSO-differentiated HL-60 cells, *Biochem. Biophys. Res. Commun.,* 145, 825, 1987.

100. **Smith, C. D., Uhing, R. J., and Snyderman, R.,** Nucleotide regulatory protein-mediated activation of phospholipase C in human polymorpho-nuclear leukocytes is disrupted by phorbol esters, *J. Biol. Chem.,* 262, 6121, 1987.

101. **O'Rourke, F., Zavoico, G. B., Smith, L. H., and Feinstein, M. B.,** Stimulus-response coupling in a cell-free platelet membrane system. GTP-dependent release of Ca^{2+} by thrombin, and inhibition by pertussis toxin-and a monoclonal antibody that blocks calcium release by IP_3. *FEBS Lett.,* 214, 176, 1987.

102. **Nakamura, T. and Ui, M.,** Simultaneous inhibitions of inositol phospholipid breakdown, arachidonic acid release, and histamine secretion in mast cells by islet-activating protein, pertussis toxin. A possible involvement of the toxin-specific substrate in the Ca^{2+}-mobilizing receptor-mediated signal transduction, *J. Biol. Chem.*, 260, 3584, 1985.

103. **Holian, A.,** Leukotriene B_4 stimulation of phosphatidylinositol turn-over in macrophages and inhibition by pertussis toxin, *FEBS Lett.*, 201, 15, 1986.

104. **Kanaide, H., Matsumoto, T., and Nakamura, M.,** Inhibition of calcium transients in cultured vasular smooth muscle cells by pertussis toxin, *Biochem. Biophys. Res. Commun.*, 140, 195, 1986.

105. **Xuan, Y. -T., Su, Y. -F., Chang, K. -J., and Watkins, W. D.,** A pertussis/cholera toxin sensitive G-protein may mediate vasopressin-induced inositol phosphate formation in smooth muscle cell, *Biochem. Biophys. Res. Commun.*, 146, 898, 1987.

106. **Bruns, C. and Marme, D.,** Pertussis toxin inhibits the angiotensin-II and serotonin-induced rise of free cytoplasmic calcium in cultured smooth muscle cells from rat aorta, *FEBS Lett.*, 212, 40, 1987.

107. **Clark, M. A., Conway, T. M., Bennett, C. F., Crooke, S. T., and Stadel, J. M.,** Islet-activating protein inhibits leukotriene D_4- and leukotriene C_4- but not bradykinin- or calcium ionophore-induced prostacyclin synthesis in bovine endothelial cells, *Proc. Natl. Acad. Sci. U.S.A.*, 83, 7320, 1986.

108. **Higashida, H., Streaty, R. A., Klee, W., and Nirenberg, M.,** Bradykinin-activated transmembrane signals are coupled via N_o or N_i to production of inositol 1, 4, 5-trisphosphate, a second messenger in NG108-15 neuroblastoma-glioma hybrid cells, *Proc. Natl. Acad. Sci. U.S.A.*, 83, 942, 1986.

109. **Shayman, J. A., Morrissey, J. J., and Morrison, A. R.,** Islet-activating protein inhibits kinin-stimulated inositol phosphate production, calcium mobilization, and prostaglandin E_2 synthesis in renal papillary collecting tubule cells independent of cyclic AMP, *J. Biol. Chem.*, 262, 17083, 1987.

110. **Backlund, P. S., Meade, B. D., Manclark, C. R., Cantoni, G. L, and Aksamit, R. R.,** Pertussis toxin inhibition of chemotaxis and the ADP-ribosylation of a membrane protein in a human mouse hybrid cell line, *Proc. Natl. Acad. Sci. U. S. A.*, 82, 2637, 1987.

111. **Pfeilschifter, J. and Bauer, C.,** Pertussis toxin abolishes angiotensin II-induced phosphoinositide hydrolysis and prostaglandin sysnthesis in rat renal mesangial cells, *Biochem. J.*, 236, 289, 1986.

112. **Stracke, M. L., Guirguis, R., Liotta, L, A., and Schiffmann, E.,** Pertussis toxin inhibits stimulated motility independently of the adenylate cyclase pathway in human melanoma cells, *Biochem. Biophys. Res. Commun.*, 146, 339, 1987.

113. **Sugiyama, H., Ito, I., and Hirono, C.,** A new type of glutamate receptor linked to inositol phospholipid metabolism, *Nature*, 325, 531, 1987.

114. **Lapetina, E. G., Reep, B., and Chang, K. -J.,** Treatment of human platelets with trypsin, thrombin, or collagen inhibits the pertussis toxin-induced ADP-ribosylation of a 41-kDa protein, *Proc. Natl. Acad. Sci. U.S.A.*, 83, 5880, 1986.

115. **Okajima, F., Katada, T., and Ui, M.,** Coupling of the guanine nucleotide regulatory protein to chemotactic peptide receptors in neutrophil membranes and its uncoupling by islet-activating protein, pertussis toxin. A possible role of the toxin substrate in Ca^{2+}- mobilizing receptor-mediated signal transduction, *J. Biol. Chem.*, 260, 6761, 1985.

116. **Ryu, S. H., Suh, P. -G., Cho, K. S., Lee, K. -Y, and Rhee, S. G.,** Bovine brain cytosol contains three immunologically distinct forms of inositol phospholipid-specific phospholipase C., *Proc. Natl. Acad. Sci. U.S.A.*, 84, 6649, 1987.

117. **Deckmyn, H., Tu, S.- M., and Majerus, P. W.,** Guanine nucleotides stimulate soluble phosphoinositde-specific phospholipase C in the absence of membranes, *J. Biol. Chem.*, 261, 16553, 1986.

118. **Wang, P., Toyoshima, S., and Osawa, T.,** Physical and functional association of cytosolic phosphatidylionsitol-specific phospholipase C of calf thymocytes with a GTP-binding protein, *J. Biochem.*, 102, 1275, 1987.

119. **Journot, L., Homburger, V., Pantaloni, C., Priam, M., Bockaert, J., and Enjalbert, A.,** An islet-activating protein-sensitive G protein is involved in dopamine inhibition of angiotensin and thyrotropin-releasing hormone-stimulated inositol phosphate production in anterior pituitary cells, *J. Biol. Chem.*, 262, 15106, 1987.

120. **Burgoyne, R. D., Cheek, T. R., and O'Sullivan, A. J.,** Receptor-activation of phospholipase A_2 in cellular signalling, *Trends Pharmacol, Sci.*, 12, 332, 1987.

121. **Burch, R. M., Luini, A., and Axelrod, J.,** Phospholipase A_2 and phospholipase C are activated by distinct GTP-binding proteins in response to α_1-adrenergic stimulation in FRTL5 thyroid cells, *J. Biol. Chem.*, 83, 7201, 1986.

122. **Nakashima, S., Tohmatsu, T., Hattori, H., Suganuma, A., and Nozawa, Y.,** Guanine nucleotides stimulate arachidonic acid release by phospholipase A_2 in saponin-permeabilized human platelets, *J. Biochem.*, 101, 1055, 1987.

123. **Okano, Y., Yamada, K., Yano, K., and Nozawa, Y.,** Guanosine 5'- [γ - thio] triphosphate stimulates arachidonic acid liberation in permeabilized rat peritoneal mast cells, *Biochem. Biophys. Res. Commun.*, 145, 1267, 1987.

124. **Murayama, T. and Ui, M.,** Receptor-mediated inhibition of adenylate cyclase and stimulation of arachidonic acid release in 3T3 fibroblasts. Selective susceptibility to islet-activating protein, pertussis, toxin, *J. Biol. Chem.,* 260, 7226, 1985.

125. **Murayama, T. and Ui, M.,** Phosphatidic acid may stimulate membrane receptors mediating adenylate cyclase inhibition and phospholipid breakdown in 3T3 fibroblasts, *J. Biol. Chem.,* 262, 5522, 1987.

126. **Fuse, I. and Tai, H. -H.,** Stimulation of arachidonate release and inositol-1,4,5-trisphosphate formation are mediated by distinct G-proteins in human platelets, *Biochem. Biophys. Res. Commun.,* 146, 659, 1987.

127. **Slivka, S. R. and Insel, P. A.,** α_1-Adrenergic receptor-mediated phosphoinositide hydrolysis and prostaglandin E$_2$ formation in madin-Darby canine kidney cells. Possible parallel activation of phospholipase C and phospholipase A$_2$, *J. Biol. Chem.,* 262, 4200, 1987.

128. **Burch, R. M. and Axelrod, J.,** Dissociation of bradykinin-induced prostaglandin formation from phosphatidylinositol turnover in Swiss 3T3 fibroblasts: evidence for G protein regulation of phospholipase A$_2$, *Proc. Natl. Acad. Sci., U.S.A.,* 84, 6374, 1987.

129. **Jelsema, C. L.,** Light activation of phospholipase A$_2$ in rod outer segments of bovine retina and its modulation by GTP-binding proteins, *J. Biol. Chem.,* 262, 163, 1987.

130. **Jelsema, C. L. and Axelrod, J.,** Stimulation of phospholipase A$_2$ activity in bovine rod outer segments by the $\beta\gamma$ subunits of transducin and its inhibition by the α subunit, *Proc. Natl. Acad. Sci. U.S.A.,* 84, 3623, 1987.

131. **Higashida, H. and Brown, D. A.,** Two polyphosphatidylinositide metabolites control two K$^+$ currents in a neuronal cell, *Nature,* 323, 333, 1986.

132. **Evans, M. G. and Marty, A.,** Potentiation of muscarinic and α-adrenergic responses by an analogue of guanosine 5'-triphosphate, *Proc. Natl. Acad. Sci. U.S.A.,* 83, 4099, 1986.

133. **Morris, A. P., Gallacher, D. V., Irvine, R. F., and Petersen, O. H.,** Synergism of inositol trisphosphate and tetrakisphosphate in activating Ca^{2+}-dependent K$^+$ channels, *Nature,* 330, 653, 1987.

134. **Aghajanian, G. K. and Wang, Y. -Y,** Pertussis toxin blocks the outward currents evoked by opiate and α_2-agonists in locus coeruleus neurons, *Brain Res.,* 371, 390, 1986.

135. **Endoh, M., Maruyama, M., and Iijima, T.,** Attenuation of muscarinic cholinergic inhibition by islet-activating protein in the heart, *Am. J. Physiol.,* 249, H309, 1985.

136. **Martin, J. M., Hunter, D. D., and Nathanson, N. M.,** Islet-activating protein inhibits physiological responses evoked by cardiac muscarinic acetylocholine receptors. Role of guanosine triphosphate binding proteins in regulation of potassium permeability, *Biochemistry,* 24, 7521, 1985.

137. **Pfaffinger, P. J., Martin, J. M., Hunter, D. D., Nathanson, N. M., and Hille, B.,** GTP-binding proteins couple cardiac muscarinic receptors to a K channel, *Nature,* 317, 536, 1985.

138. **Sorota, S., Tsuji, Y., Tajima, T., and Pappano, A. J.,** Pertussis toxin treatment blocks hyperpolarization by muscarinic agonists in chick atrium, *Circ. Res,* 57, 748, 1985.

139. **Breitwieser, G. E. and Szabo, G.,** Uncoupling of cardiac muscarinic and β-adrenergic receptors from ion channels by a guanine nucleotide analogue, *Nature,* 317, 538, 1985.

140. **Kurachi, Y., Nakajima, T., and Sugimoto, T.,** On the mechanism of activation of muscarinic K$^+$ channels by adenosine in isolated atrial cells: involvement of GTP-binding proteins, *Pflugers Arch.,* 407, 264, 1986.

141. **Sasaki, K. and Sato, M.,** A single GTP-binding protein regulates K$^+$-channels coupled with dopamine, histamine and acetylcholine receptors, *Nature,* 325, 259, 1987.

142. **Andrade, R., Malenka, R. C., and Nicoll, R. A.,** A G protein couples serotonin and GABA$_B$ receptors to the same channels in hippocampus, *Science* 234, 1261, 1986,.

143. **Miwa, A., Kawai, N., and Ui, M.,** Pertussis toxin blocks presynaptic glutamate receptors—a novel 'glutamate$_B$' receptor in the lobster neuromuscular synapse, *Brain Res.,* 416, 162, 1987.

144. **Yatani, A., Codina, J., Sekura, R. D., Birnbaumer, L., and Brown, A. M.,** Reconstitution of somatostatin and muscarinic receptor mediated stimulation of K$^+$ channels by isolated G$_k$ protein in clonal rat anterior pituitary cell membranes, *Mol. Endocrinol.,* 1, 283, 1987.

145. **North, R. A., Williams, J. T., Surprenant, A., and Christie, M. J.,** μ and δ receptors belong to a family of receptors that are coupled to potassium channels, *Proc. Natl. Acad. Sci. U.S.A.,* 84, 5487, 1987.

146. **Piomelli, D., Volterra, A., Dale, N., Siegelbaum, S. A., Kandel, E. R., Schwartz, J. H., and Belardetti, F.,** Lipoxygenase metabolites of arachidonic acid as second messengers for presynaptic inhibition of *Aplysia* sensory cells, *Nature,* 328, 38, 1987.

147. **Codina, J., Grenet, D., Yatani, A., Birnbaumer, L, and Brown, A. M.,** Hormonal regulation of pituitary GH$_3$ cell K$^+$ channels by G$_k$ is mediated by its α-subunit, *FEBS Lett.,* 216, 104, 1987.

148. **Logothetis, D. E., Kurachi, Y., Galper, J., Neer, E. J., and Clapham, D. E.,** The $\beta\gamma$ subunits of GTP-binding proteins activate the muscarinic K$^+$ channel in heart, *Nature,* 325, 321, 1987.

149. **Bourne, H. R.,** 'Wrong' subunit regulates cardiac potassium channels, *Nature,* 325, 296, 1987.

150. **Birnbaumer, L.,** Which G protein subunits are the active mediators in signal transduction?, *Trends Pharmacol. Sci.,* 8, 209, 1987.

151. **Hofman, F., Nastainczyk, W., Rohrkasten, A., Schneider, T., and Sieber, M.,** Regulation of the L-type calcium channel, *Trends Pharmacol, Sci.,* 8, 393, 1987.

152. **Rosenthal, W. and Schultz, G.,** Modulations of voltage-dependent ion channels by extracellular signals, *Trends Pharmacol. Sci.,* 8, 351, 1987.

153. **Nakamura, T. and Gold, G. H.,** A cyclic nucleotide-gated conductance in olfactory receptor cilia, *Nature,* 325, 442, 1987.

154. **Lancet D. and Pace. U.,** The molecular basis of odor recognition, *Trends Biochem. Sci.,* 12, 63, 1987.

155. **Kojima, I., Shibata, H., and Ogata, E.,** Pertussis toxin blocks angiotensin II-induced calcium influx but not inositol trisphosphate production in adrenal glomerulasa cell, *FEBS Lett.,* 204, 347, 1986.

156. **Nomura, Y. and Tohda, M.,** Inhibitory effects of pertussis toxin on a depolarization-evoked Ca^{2+} influx in NG108-15 cells, *FEBS Lett.,* 216, 40, 1987.

157. **Nasmith, P. E. and Grinstein, S.,** Phorbol ester-induced changes in cytoplasmic Ca^{2+} in human neutrophils. Involvement of a pertussis toxin-sensitive G protein, *J. Biol. Chem.,* 262, 13558, 1987.

158. **Yatani, A., Codina, J., Imoto, Y., Reeves, J. P., Birnbaumer, L., and Brown, A. M.,** A G protein directly regulates mammalian cardiac calcium channels, *Science,* 238, 1288, 1987.

159. **Scott, R. H. and Dolphin, A. C.,** Activation of a G protein promotes agonist responses to calcium channel ligands, *Nature,* 330, 760, 1987.

160. **Heisler, S.,** The inhibitory guanine nucleotide-binding regulatory subunit of adenylate cyclase has an adenylate cyclase-independent modulatory effect on ACTH secretion from mouse pituitary tumor cells, *Biochem. Biophys. Res. Commun.,* 126, 941, 1985.

161. **Schlegel, W., Wuarin, F., Zbaren, C., Wollheim, C. B., and Zahnd, G. R.,** Pertussis toxin selectively abolishes hormone induced lowering of cytosolic calcium in GH_3 cells, *FEBS Lett.,* 189, 27, 1985.

162. **Koch, B. D., Dorflinger, L. J., and Schonbrunn, A.,** Pertussis toxin blocks both cyclic AMP-mediated and cyclic AMP-independent actions of somatostatin, *J. Biol. Chem.,* 260, 13138, 1985.

163. **Reisine, T. and Guild, S.,** Pertussis toxin blocks somatostatin inhibition of calcium mobilization and reduces the affinity of somatostatin receptors for agonists, *J. Pharmacol. Exp. Ther.,* 235, 551, 1985.

164. **Yajima, Y., Akita, Y., and Saito, T.,** Pertussis toxin blocks the inhibitory effects of somatostatin on cAMP-dependent vasoactive intestinal peptide and cAMP-independent thyrotropin releasing hormone-stimulated prolactin secretion of GH_3 cells, *J. Biol. Chem.,* 261, 2684, 1986.

165. **Lewis, D. L., Weight, F. F., and Luini, A.,** A guanine nucleotide-binding protein mediates the inhibition of voltage-dependent calcium current by somatostatin in a pituitary cell line, *Proc. Natl, Acad. Sci. U.S.A.,* 83, 9035, 1986.

166. **Holz, G. G., Rane, S. G., and Dunlap, K.,** GTP-binding proteins mediate transmitter inhibition of voltage-dependent calcium channels, *Nature,* 319, 670, 1986.

167. **Malgaroli, A., Vallar, L., Elahi, F. R., Pozzan, T., Spada, A., and Meldolesi, J.,** Dopamine inhibits cytosolic Ca^{2+} increases in rat lactotroph cells, *J. Biol. Chem.,* 262, 13920, 1987.

168. **Wanke, E., Ferroni, A., Malgaroli, A., Ambrosini, A., Pozzan, T., and Meldolesi, J.,** Activation of a muscarinic receptor selectively inhibits a rapidly inactivated Ca^{2+} current in rat sympathetic neurons, *Proc. Natl. Acad. Sci. U.S.A.,* 84, 4313, 1987.

169. **Hescheler, J., Rosenthal, W., Trautwein, W., and Schultz, G.,** The GTP-binding protein, G_o, regulates neuronal calcium channels, *Nature,* 325, 445, 1987.

170. **Dolphin, A. C. and Prestwich, S. A.,** Pertussis toxin reverses adenosine inhibition of neuronal glutamate release, *Nature,* 316, 148, 1985.

171. **Crain, S. M., Crain, B., and Makman, M. H.,** Pertussis toxin blocks depressant effects of opioid, monoaminergic and muscarinic agonists on dorsal-horn network responses in spinal cord-ganglion culture, *Brain Res.,* 400, 185, 1987.

172. **Clarke, W. P., De Vivo, M., Beck, S. G., Maayani, S., and Goldfarb, J.,** Serotonin decreases population spike amplitude in hippocampal cells through a pertussis toxin substrate, *Brain Res.,* 410, 357, 1987.

173. **Innis, R. B. and Aghajanian, G. K.,** Pertussis toxin blocks autoreceptor-mediated inhibition of dopaminergic neurons in rat substantia nigra, *Brain Res.,* 411, 139, 1987.

174. **Steinberg, S. F., Drugge, E. D., Bilezikian, J. P., and Robinson, R. B.,** Acquisition by innervated cardiac myocytes of a pertussis toxin specific regulatory protein linked to the α_1-receptor, *Science,* 230, 186, 1985.

175. **Hildebrandt, J. D., Stolzenberg, E., and Graves, J.,** Pertussix toxin alters the growth characteristics of Swiss 3T3 cells, *FEBS Lett.,* 203, 87, 1986.

176. **Paris, S. and Pouyssegur, J.,** Pertussis toxin inhibits thrombin-induced activation of phosphoinositide hydrolysis and Na^+/H^+ exchange in hamster fibroblasts, *EMBO J.,* 5, 55, 1986.

177. **Letterio, J. J., Coughlin, S. R., and Williams, L. T.,** Pertussis toxin-sensitive pathway in the stimulation of c-*myc* expression and DNA synthesis by bombesin, *Science,* 234, 1117, 1986.

178. **Paris, S., Chambard, J. -C., and Pouyssegur, J.,** Coupling between phosphoinositide breakdown and early mitogenic events in firbroblasts. Studies with fluoraluminate and pertussis toxin, *J. Biol. Chem.,* 262, 1977, 1987.

179. **Chambare, J. C., Paris, S., L'Allemain, G., and Pouyssegur, J.,** Two growth factor signalling pathways in fibroblasts distinguished by pertussis toxin, *Nature,* 326, 800, 1987.

180. **Magnaldo, I., L'Allemain, G., Chambard, J. C., Moenner, M., Barritault, D., and Pouyssegur, J.,** The mitogenic signaling pathway of fibroblast growth factor is not mediated through polyphosphoinositide hydrolysis and protein kinase C activation in hamster fibroblasts, *J. Biol. Chem.*, 261, 16916, 1986.

181. **Paris, S. and Pouyssegur, J.,** Further evidence for a phospholipase C-coupled G protein in hamster fibroblasts. Induction of inositol phosphate formation by fluoroaluminate and vanadate and inhibition by pertussis toxin, *J. Biol. Chem.*, 262, 1970, 1987.

182. **Zachary, I., Millar, J., Nanberg, E., Higgins, T., and Rozengurt, E.,** Inhibition of bombesin-induced mitogenesis by pertussis toxin: dissociation from phospholipase C pathway, *Biochem. Biophys. Res. Commun.*, 146, 456, 1987.

183. **Taylor, C. W., Blakeley, D. M., Corps, A. N., Berridge, M. J., and Brown, K. D.,** Effects of pertussis toxin on growth factor-stimulated inositol phosphate formation and DNA systhesis in Swiss 3T3 cells, *Biochem. J.*, 249, 917, 1988.

184. **Burch, R. M., Luini, A., Mais, D. E., Corda, D., Vanderhoek, J. Y., Kohn, L. D., and Axelrod, J.,** α_1-Adrenergic stimulation of arachidonic acid release and metabolism in a rat thyroid cell line, *J. Biol. Chem.*, 261, 11236, 1986.

185. **Okajima, F., Tokumitsu, Y., Kondo, Y., and Ui, M.,** P_2-Purinergic receptors are coupled to two signal transduction systems leading to inhibition of cAMP generation and to production of inositol trisphosphate in rat hepatocytes, *J. Biol. Chem.*, 262, 13483, 1987.

186. **Johnson, R. M., Connelly, P. A., Sisk, R. B., Pobiner, B. F., Hewlett, E. L., and Garrison, J. C.,** Pertussis toxin or phorbol 12-myristate 13-acetate can distinguish between epidermal growth factor- and angiotensin-stimulated signals in hepatocytes, *Proc. Natl. Acad. Sci. U.S.A.*, 83, 2032, 1986.

187. **Johnson, R. M. and Garrison, J. C.** Epidermal growth factor and angiotensin II stimulate formation of inositol 1,4,5- and inositol 1,3,4-trisphosphate in hepatocytes. Differential inhibiton by pertussis toxin and phorbol 12-mypistate and 13-acetate, *J. Biol. Chem.*, 262, 17285, 1987.

188. **Hughes, B. P. and Barritt, G. J.,** The stimulation by sodium fluoride of plasma-membrane Ca^{2+} inflow in isolated hepatocytes. Evidence that a GTP-binding protein is involved in the hormonal stimulation of Ca^{2+} inflow, *Biochem., J.*, 245, 41, 1987.

189. **Hughes, B.P., Crofts. J. N., Auld, A. M., Read, L. C., and Barritt, G. J.,** Evidence that a pertussis-toxin-sensitive substrate is involved in the stimulation by epidermal growth factor and vasopressin of plasma-membrane Ca^{2+} inflow in hepatocytes, *Biochem. J.*, 248, 911, 1987.

190. **Nishimoto, I., Ohkuni, Y., Ogata, E., and Kojima, I.,** Insulin-like growth factor II increases cytoplasmic free calcium in competent Balb/c 3T3 cells treated with epidermal growth factor, *Biochem. Biophys. Res. Commun.*, 142, 275, 1987.

191. **Nishimoto, I., Hata, y., Ogata, E., and Kojima, I.,** Insulin-like growth factor II stimulates calcium influx in competent BALB/c 3T3 cells primed with epidermal growth factor, *J. Biol. Chem.*, 262, 12120, 1987.

192. **Nishimoto, I., Ogata, E., and Kojima, I.,** Pertussis toxin inhibits the action of insulin-like growth factor-I, *Biochem. Biophys. Res. Commun.*, 148, 403, 1987.

193. **Murayama, T. and Ui, M.,** Possible involvement of a GTP-binding protein, the substrate of islet-activating protein, in receptor-mediated signaling responsible for cell proliferation, *J. Biol. Chem.*, 262, 12463, 1987.

194. **Pouyssegur, J.,** The growth factor-activatable Na^+/H^+ exchange system: a genetic approach, *Trends Biochem. Sci.*, 6, 453, 1985.

195. **Moolenaar, W. H.,** Effects of growth factors on intracellular pH regulation, *Annu. Rev. Physiol.*, 48, 363, 1986.

196. **Busa, W. B.,** Mechanisms and consequences of pH-mediated cell regulation, *Annu. Rev. Physiol.*, 48, 389, 1986.

197. **Sweatt, J. D., Blair, I. A., Cragoe, E. J., and Limbird, L. E.,** Inhibitors of Na^+/H^+ exchange block epinephrine- and ADP-induced stimulation of human platelet phospholipase C by blockade of arachidonic acid release at a prior step, *J. Biol. Chem.*, 261, 8660, 1986.

198. **Sweatt, J. D., Connolly, T. M., Cragoe, E. J., and Limbird, L. E.,** Evidence that Na^+/H^+ exchange regulates receptor-mediated phospholipase A_2 activation in human platelets, *J. Biol. Chem.*, 261, 8667, 1986.

199. **Banga, H. S., Simons, E. R., Brass, L. F., and Rittenhouse, S. E.,** Activation of phospholipase A and C in human platelets exposed to epinephrine: role of glycoproteins IIb/IIIa and dual role of epinephrine, *Proc. Natl. Acad. Sci. U.S.A.*, 83, 9197, 1986.

200. **Isom, L. L., Cragoe, E. J., and Limbird, L. E.,** α_2-Adrenergic receptors accelerate Na^+/H^+ exchange in neuroblastoma × glioma cells, *J. Biol. Chem.*, 262, 6750, 1987.

201. **Repaske, M. G., Nunnari, J. M., and Limbird, L. E.,** Purification of the α_2-adrenergic receptor from porcine brain using a yohimibine-agarose affinity matrix, *J. Biol. Chem.*, 262, 12381, 1987.

202. **Nunnari, J. M., Repaske, M. G., Brandon, S., Cragoe, E. J., and Limbird, L. E.,** Regulation of porcine brain α_2-adrenergic receptors by Na^+ H^+ and inhibitors of Na^+/H^+ exchange, *J. Biol. Chem.*, 262, 12387, 1987.

203. **Isom, L. L., Cragoe, E. J., and Limbird, L. E.,** Multiple receptors linked to inhibition of adenylate cyclase accelerate Na^+/H^+ exchange in neuroblastoma $\times$ glioma cells via a mechanism other than decreased cAMP accumulation, *J. Biol. Chem.,* 262, 17504, 1987.

204. **Huang, C.-L, Cogan, M. G., Cragoe, E. J., and Ives, H. E.,** Thrombin activation of the Na^+H^+ exchanger in vascular smooth muscle cells, *J. Biol. Chem.,* 262, 14134, 1987.

205. **Rozengurt, E.,** Early signals in the mitogenic response, *Science,* 234, 161, 1986.

206. **Rozengurt, E., Murray, M., Zachary, I., and Collins, M.,** Protein kinase C activation enhances cAMP accumulation in Swiss 3T3 cells: inhibition by pertussis toxin, *Proc. Natl. Acad. Sci. U.S.A.,* 84, 2282, 1987.

207. **Pirotton, S., Erneux, C., and Boeynaems, J. M.,** Dual role of GTP-binding proteins in the control of endothelial prostacyclin, *Biochem. Biophys. Res. Commun.,* 147, 1113, 1987.

208. **Bianca, V. D., Grzeskowiak, M., Cassatella, M. A., Zeni, L., and Rossi, F.,** Phorbol 12, myristate 13, acetate potentiates the respiratory burst while inhibits phosphoinositide hydrolysis and calcium mobilization by formyl-methionyl-leucyl-phenylalanine in human neutrophils, *Biochem. Biophys. Res. Commun.,* 135, 556, 1986.

209. **Grzeskowiak, M., Bianca, V. D., Cassatella, M., A., and Rossi, F.,** Complete dissociation between the activation of phosphoinositide turnover and of NADPH oxidase by formyl-methionyl-leucyl-phenylalanine in human neutrophils depleted of Ca_2^+ and primed by subthreshold doses of phorbol 12, myristate 13, acetate, *Biochem. Biophys. Res. Commun.,* 135, 785, 1986.

210. **Grinstein, S. and Furuya, W.,** Receptor-mediated activation of electropermeabilized neutrophils. Evidence for a Ca^{2+} and protein kinase C-independent signaling pathway, *J. Biol. Chem.* 263, 1779, 1988.

211. **Saito, H., Okajima, F., Molski, T. F. P., Sha'afi, R. I., Ui, M., and Ishizaka, T.,** Effects of ADP-ribosylation of GTP-binding protein by pertussis toxin on immunogloubulin E-dependent and -independent histamine release from mast cells and basophils, *J. Immunol.,* 138, 3927, 1987.

212. **Barrowman, M. M. Cockcroft, S., and Gomperts, B. D.,** Two roles for guanine nucleotides in the stimulus-secretion sequence of neutrophils, *Nature,* 319, 504, 1986.

213. **Bittner, M. A., Holz, R. W., and Neubig, R. R.,** Guanine nucleotide effects on secretion from digitonin-permeabilized adrenal chromaffin cells, *J. Biol. Chem.,* 261, 10182, 1986.

214. **Vallar, L., Biden, T. J., and Wollheim, C. B.,** Guanine nucleotides induce Ca^2-independent insulin secretion from permeabilized RINm5F cells, *J. Biol. Chem.,* 262, 5049, 1987.

215. **Johnson, J. D. and Davies, P. J. A.,** Pertussis toxin inhibits retinoic acid-induced expression of tissue transglutaminase in macrophages, *J. Biol. Chem.,* 261, 14982, 1986.

216. **Hughes, A. R., Martin, M. W., and Harden, T. K.,** Pertussis toxin differentiates between two mechanisms of attenuation of cyclic AMP accumulation by muscarinic cholinergic receptors, *Proc. Natl. Acad. Sci. U.S.A.,* 81, 5680, 1984.

217. **Gutierrez, G. R., Derynck, R., Hewlett, E. L., and Katz. M. S.,** Inhibition of parathyroid hormone-responsive adenylate cyclase in clonal osteoblast-like cells by transforming growth factor α and epidermal growth factor, *J. Biol. Chem.,* 262, 15845, 1987.

218. **Kikuchi, A., Yamashita, T., Kawata, M., Yamamoto, K., Ikeda, K., Tanimoto, T., and Takai, Y.,** Purification and characterization of a novel GTP-binding protein with a molecular weight of 24,000 from bovine brain membranes, *J. Biol. Chem.,* 263, 2897, 1988.

219. **Katada, T., Imai, S., Tohkin, M., and Ui, M.,** Purification and characterization of a new GTP-binding protein with the molecular mass of 24,000 daltons from porcine brain membranes, *J. Biol. Chem.,* 263, in press.

220. **Brown, J. H. and Brown, S. L.,** Agonists differentiate muscarinic receptors that inhibit cyclic AMP formation from those that stimulate phosphoinositide metabolism, *J. Biol. Chem.,* 259, 3777, 1984.

221. **Ashkenazi, A., Winslow, J. W., Peralta, E. G., Peterson, G. L., Schimerlik, M. I., Capon, D. J., and Ramachandran, J.,** An M_2 muscarinic receptor subtype coupled to both adenylyl cyclase and phosphoinositide turnover, *Science,* 238, 672, 1987.

Chapter 2

ONCOGENE PRODUCTS INVOLVED IN SIGNAL TRANSDUCTION

Alan Hall

Table of Contents

I.   Introduction 30

II.   Growth Factor Receptor Oncogenes... 30
        A.   Introduction 30
        B.   Epidermal Growth Factor (EGF) Receptor (c-*erb*B1)..................... 31
        C.   c-*erb*B2 or c-*neu* 31
        D.   c-*fms* 32
        E.   Conclusions 32

III.   Intracellular Tyrosine Kinases 32

IV.   *Ras* Proteins 33
        A.   Structure.. 33
        B.   Biochemical Activity 34
        C.   Mutational Analysis 36
        D.   Function... 36
                1.   In yeast 37
                2.   In mammalian cells 37
        E.   *ras* Related Genes 38

V.   Conclusion... 38

References... 39

I. INTRODUCTION

Over the last decade or so a great deal has been learned about the types of changes that are involved in the development of human cancer, and it is now widely accepted that a malignant cell arises as a result of an accumulation of specific genetic alterations. Well over 50 genes have been identified in the human genome as potential sites for oncogenic lesions, the so-called protooncogenes, though for only a handful of these is there yet any direct evidence of involvement in human malignancies (see reference 1 for review).

Many of the protooncogenes and their protein products have been studied in detail in an attempt to understand the biochemical changes associated with malignant transformation. It has become apparent that most protooncogenes described so far code for proteins involved in controlling cell growth,[2] and examples of proteins located in the nucleus (e.g., the *myc* family, *myb,* and *fos*), in the cytoplasm, (e.g., *raf* and *mos*), and in the plasma membrane (e.g., *erb*B, *ras, src*) are all known. There is growing evidence that they each form part of the normal cascade of events leading from an extracellular mitogenic signal through to a nuclear signal for DNA synthesis and cell division. An alteration in any one of these protooncogenes that leads either to uncontrolled or ectopic expression or to the expression of a qualitatively different product might be expected to lead to a breakdown in the regulation of cell growth.

This chapter will deal with oncogenes that code for proteins located at the plasma membrane and which are thought to be involved with the early events of mitogenic signal transduction. These proteins include transmembrane growth factor receptors, tyrosine kinases, and guanine nucleotide regulatory proteins (G proteins). Particular attention will be given to the *ras* gene products which are located on the inside of the plasma membrane and are thought to be G proteins that play an essential regulatory role in cell proliferation. Approximately 30% of human cancers have a *ras* gene containing an activating point mutation and an understanding of the biochemical function of this protein will not only shed light on the mechanism of normal growth control but may also go some way to explain the abnormal behaviour of a large proportion of human malignancies.

II. GROWTH FACTOR RECEPTOR ONCOGENES

A. INTRODUCTION

Growth factor receptors are transmembrane molecules with an extracellular ligand binding site linked via a single hydrophobic transmembrane region to an intracellular domain. It is assumed that binding of ligand activates the intracellular domain but it is not known how this signal is interpreted by the cell. In the case of hormone receptors that affect cAMP levels, the first intracellular interaction is between the stimulated receptor and a G protein.[3] This leads to activation of the G protein which in turn can interact directly with adenylate cyclase. Some growth factor receptors appear to work through a G protein in a similar way and affect cAMP levels, e.g., PGE_1, in Swiss 3T3 fibroblasts.[4]

Many other receptors, however, (e.g., those for $PGF_{2\alpha}$, vasopressin, bombesin, and PDGF) do not act directly through adenylate cyclase but instead can activate a phospholipase C,[5-7] which hydrolyses phosphatidylinositol 4,5 bisphosphate (PIP_2) to produce two second messengers, 1,2 diacylglycerol (DAG) and 1,4,5 inositol trisphosphate (IP_3).[8] There is growing evidence for the involvement of at least two distinct G proteins in the phospholipase C activation process, one pertussis toxin sensitive and one insensitive, but it is not clear whether they are activated directly by receptors.[9,10]

Unlike hormonal receptors that couple to adenylate cyclase through stimulatory (G_s) or inhibitory (G_i) G proteins, many growth factor receptors have a tyrosine kinase activity encoded within their intracellular domain. This is true of receptors that function through the phospholipase C pathway (e.g., PDGF receptor) as well as for receptors that do not appear to activate directly either adenylate cyclase or phospholipase C (e.g., EGF receptor, CSF1 receptor, and the insulin receptor). It is likely therefore that a phosphorylation event is the first signal generated by a stimulated tyrosine kinase receptor.

Several examples are known where structural alterations in a growth factor receptor gene can lead to cell transformation, i.e., the altered receptor gene acts as an oncogene. An analysis of the effects of these oncogenes on cells has given some insight into the ways in which the normal receptors function.

B. EPIDERMAL GROWTH FACTOR (EGF) RECEPTOR (c-*erb*B1)

The oncogene v-*erb*B is responsible for the transforming properties of the avian erythroblastosis virus. Sequence analysis has shown that this oncogene is a truncated form of the cellular EGF receptor gene.[11] One obvious feature of the v-*erb*B protein is that the extracellular domain of the receptor has been deleted and it was originally supposed that this leads to growth factor independent, i.e., constitutively activated receptor. In agreement with this, it had been shown that in chicken erythroblasts simple amino-terminal truncation of the endogenous c-*erb*1 gene by viral integration was sufficient to generate an oncogene.[12] Recent work, however, has suggested that it may not be so simple since fusion of an intact extracellular domain onto v-*erb*B does not eliminate the transforming ability of the gene when introduced into Rat-1 fibroblasts. It appears that the small (32 amino acid) carboxy-terminal deletion is crucial for transformation,[13] and perhaps this alters the substrate specificity or the regulation of the kinase activity. An explanation for these apparently contradictory results may lie in the different cell types (erythroblasts and fibroblasts) used for assaying transformation.

The kinase activity of the receptor is essential and mutations that destroy it generate nonfunctional receptors.[14] Addition of ligand causes autophosphorylation at three sites within the cytoplasmic domain but the effect of this on the catalytic activity is unclear.[15,16] Interestingly, the main autophosphorylation site is removed by the carboxy terminal deletion in v-*erb*B.

The signal pathways activated by the EGF receptor are unknown. In one cell line (A431) which contains very high levels of the EGF receptor, addition of EGF stimulates phospholipase C activity but this has not been observed in other cells.[17] It has been reported that EGF stimulates phosphorylation of a 35 kDa protein which has been identified as lipocortin I, an inhibitor or phospholipase A_2.[18,19] Phosphorylation of this protein on tyrosine is thought to block its inhibitory action and it is possible that activation of the EGF receptor might thereby activate PLA_2.

C. c-*erb*B2 OR c-*neu*

Sequence analysis of an activated oncogene identified in chemically induced rat neuroblastomas, originally called c-*neu*,[21] shows that it is very similar to c-*erb*B1, the EGF receptor gene. The protein from rat c-*neu*, (called c-*erb*B2 in humans), has a predicted tyrosine kinase domain but does not bind EGF, and its ligand is unknown.[21] What is particularly interesting is a comparison between the normal and activated *neu* genes. The only difference is a single amino acid substitution, Val to Glu in the transmembrane hydrophobic domain.[22] This single amino acid substitution in the 185 kDa protein is sufficient to generate an oncoprotein. The location of the mutation suggests an alteration in the interaction between the external and internal domains.

There are two current ideas for how the two domains of transmembrane receptors might interact. It is possible that ligand binding induces a conformational change in the intracellular domain via the transmembrane sequence. Perhaps this is mimicked in activated c-*neu* by the amino acid substitution. An alternative hypothesis is that ligand binding induces oligomerization which, in the case of the EGF receptor at least, is thought to activate the kinase activity.[23] It is possible that the change in activated c-*neu* could favor this association perhaps even in the absence of ligand.

D. c-*fms*

The oncogene carried by the McDonough strain of feline sarcoma virus, v-*fms*, is derived from the cellular c-*fms* gene which encodes the macrophage colony stimulating factory (CSF1) receptor.[24] Although v-*fms* has an intact ligand binding domain, the kinase is constitutively activated and no longer requires ligand stimulation.[25] The reasons for this are still not entirely clear. In addition to amino acid alterations within the body of the protein, 40 amino acids at the C-teminus of the normal receptor are replaced by 11 unrelated amino acids in the oncogene. This removes a tyrosine residue which is normally phosphorylated. Recent experiments have shown that phosphorylation of this tyrosine negatively regulates the kinase activity and hence its removal contributes to kinase activation.[26] The substrate for the CSF-1 receptor is not known but activation of PI turnover does not seem to be involved, at least in macrophages.[27]

E. CONCLUSIONS

Perhaps, surprisingly, no viral oncogene derived from the PDGF or the insulin receptor has so far been identified, even though in the case of PDGF the gene for the growth factor itself can function as an oncogene (v-*sis*). The genes for these two receptors have now been cloned[28,29] and already mutational analysis is under way. It has been shown that *in vitro* mutagenesis of the insulin receptor gene can generate an oncogene with biological effects on NIH-3T3 cells.[30] It is also possible that mutational analysis might resolve whether the two activities associated with the PDGF receptor, namely a tyrosine kinase and stimulation of phospholipase C activity, are connected or represent different signaling systems.

Biological screening has identified several oncogenes encoding receptor-like proteins whose function is still not known. c-*neu* has already been discussed. The v-*ros* and v-*kit* oncogenes have many structural similarities to the insulin and PDGF receptors, respectively, but their normal counterparts are unknown.[29,31] A different kind of receptor-derived oncogene, c-*mas*, has been detected using the NIH-3T3 transfection assay.[32] Sequence analysis of this gene suggest that it too codes for a receptor though its predicted structure resembles rhodopsin or the β adrenergic receptor with multiple potential transmembrane domains rather than a typical growth factor receptor with a single transmembrane domain.

Although mutational analysis and comparisons between receptors and oncogenic derivatives have clearly enabled essential regions and activities of the molecules to be defined, what is still lacking is hard information concerning the intracellular substrate for activated growth factor receptors. Of the possible candidates that have been proposed, two groups of proteins, tyrosine kinases and G proteins, have been found to function independently as oncogenes.

III. INTRACELLULAR TYROSINE KINASES

At least five oncoproteins have been characterized which encode intracellular tyrosine kinase activities localized on the plasma membrane.[2] The best studied example is v-*src*, the oncogene carried by the Rous sarcoma virus. This viral gene is derived from c-*src* and both encode phosphorylated 60,000 dalton proteins (pp60$^{v\text{-}src}$, pp60$^{c\text{-}src}$ which are attached to the plasma membrane via amino-terminal myristylation. In addition to the 19 C-terminal amino acids of c-*src* being replaced with 12 different amino acids in v-*src*, there are 8 internal amino acid differences.[33] Simple overexpression of c-*src* in cells does not lead to cell transformation and at least some of the amino acid substitutions must be functionally significant.[34] Recent mutational analysis has suggested that phosphorylation of tyr 527 is a negative regulator of the tyrosine kinase activity,[35] and mutation of this amino acid in c-*src* can give rise to a transforming gene. This story is still more complicated since removal of tyr 527 causes a large increase in phosphorylation of tyr 416. If this residue is mutated then this blocks the transforming activity of the tyr 527 mutant. It looks as though the kinase activity of pp60$^{c\text{-}src}$ is down-regulated *in vivo* by phosphorylation, but this inhibitory action is blocked in pp60$^{v\text{-}src}$ by a combination of a C-terminal deletion and point mutations, leading to a constitutively active kinase. This is very reminiscent of the emerging story of how the EGF and CSF-1 receptors can be activated (see Section II).

A similar picture looks likely for the control of the *abl* gene protein kinase. Although the proteins from v-*abl* (the oncogene of the Abelson murine leukemia virus) and the protooncogene c-*abl* both have tyrosine kinase activities, their sites of autophosphorylation are different.[36] In this respect the product of the translocated c-*abl* gene present in Philadelphia chromosome positive human chronic myelogenous leukemias resembles v-*abl*, and they both have greatly enhanced kinase activity compared to normal c-*abl*.

It is still unclear where c-*sec*, c-*abl*, and other members of the tyrosine kinase family fit into the signal transduction process. Many proteins are phosphorylated by pp60^{src} but it seems unlikely that any so far identified are the primary target.[37] Mutants of pp60$^{v\text{-}src}$ that cannot be myristylated phosphorylate the same group of proteins, even though these mutants do not lead to cell transformation.[38] Perhaps the substrate required for transformation is membrane bound and/or inaccessible to the soluble kinase. More recently, it has been shown that although nonmyristylated pp60$^{v\text{-}src}$ does not transform cells it will stimulate cell proliferation,[39] raising the possibility that there is more than one substrate. One cellular substrate for pp60^{src}, a 36 kDa protein, has been studied in some detail. The cDNA has been cloned and the protein is 50% homologous to lipocortin I.[40] The protein, called lipocortin II or calpactin I, is an inhibitor of phospholipase A_2.

There was a great deal of excitement in 1984 when Sugimoto et al. and Macara et al. showed that a preparation of v-*src* or v-*ros* protein could phosphorylate phosphatidylinositol.[41,42] It now seems that the phosphorylation was due to a contaminating protein.[43] However, a PI kinase activity has been detected in immunoprecipitates of polyoma middle T/c-*src* complexes.[44-46] It is an exciting possibility, therefore, that the kinases affecting the levels of PI, PIP, and PIP_2 in cells are closely associated with and perhaps even phosphorylated by pp60^{src}. Confirmation of this will have to await purification of the lipid kinases involved.

IV. *RAS* PROTEINS

Mammalian genomes contain three functional *ras* genes, c-Ha-*ras*1, c-Ki-*ras*2, and N-*ras* (for a review see Reference 47). Each encode 21,000 dalton (p21) proteins which are expressed in almost all cells at low levels. The *ras* genes have become the focus of a great deal of attention since it was discovered that approximately 30% of human tumors contain an activated *ras* gene.[48-50] The mechanism of activation in human tumors is by a somatic point mutation altering a single amino acid either at position 12, 13, or 61.[50,51]

The effect of these mutations can be examined experimentally *in vivo* by introducing either the gene or the purified proteins into established rodent fibroblasts.[48,49,52] For example, introduction of a cloned mutant *ras* gene into NIH-3T3 cells converts the cells to a fully malignant phenotype, and the levels of mutant protein required are very low (less than 50,000 molecules/cell).[53] It has also been found that overexpression of the normal protein (10^6 to 10^7 molecules/cell) leads to transformation.[54,55]

A. STRUCTURE

p21$^{N\text{-}ras}$ and p21$^{Ha\text{-}ras}$ are 189 amino acid proteins, while p21$^{Ki\text{-}ras}$ contains 188 amino acids.[47] The Ki-*ras* gene has two possible alternative carboxy termini (encoded in exons 4A and 4B) but in all cells so far examined only 4B is expressed.[56] There are a maximum of 15 amino acid differences between any two of the proteins within the first 164 amino acids. Of the 15, 8 are clustered between amino acids 121 to 135 and most of these changes are conservative. However, amino acids 165 to 185 encode a highly variable region with no homology between the three *ras* proteins.

The proteins bind GTP and GDP and have sequence homology to regulatory guanine nucleotide binding proteins including G_s, G_i, and *E. coli* elongation factor EF-Tu.[57,58] The crystal structure of EF-Tu has been determined and a structural model for p21^{ras} has been proposed by comparison.[58] The majority of the *ras* protein appears to be involved in forming the guanine nucleotide binding pocket and Figure 1 shows a schematic representation of the proposed functional regions. Amino acids 10 to 16 and 57 to 63 are close to the phosphoryl binding region and it was proposed that 57 and 63 may also be the Mg^{2+} binding site. Amino acids Asn-116 and Asp-119 are predicted to define the guanine nucleotide specificity and this has been confirmed by mutagenesis (see Figure 1 and later).[59,60] The region 30 to 40 is expected to loop out of the structure and may be the effector region, i.e., the part of *ras* which interacts with a downstream molecule. In accordance with this, point mutations and deletions within this region as well as between amino acids 21 to 30 inactivate the biological activity of the protein without affecting guanine nucleotide binding or intrinsic GTPase activity.[61,62] By analogy with EF-Tu and other G proteins (G_s and G_i), the carboxy-terminus would be predicted to contain the detector region and interact with an upstream protein. If this assignment of effector/detector regions is correct, then the prediction would be that the three *ras* proteins are activated by different detector molecules but that they feed into the same effector pathway.

The protein is palmitoylated on cys-186 which localizes p21^{ras} to the plasma membrane, and mutation of cys-186 to ser-186 destroys the biological activity.[63] Interestingly, although the protein itself is long-lived *in vivo* ($T_{1/2}$ ~24 h), the half-life of the palmitic acid is very short ($T_{1/2}$ 20′).[64] The biological significance of this high turnover rate is not known.

Ras genes have been detected in all eukaryotic species including yeast. *Saccharomyces cerevisiae* contains two *ras* genes, RAS1 and RAS2.[65] The proteins encoded by these genes are longer (309 and 322 amino acids) with the extra amino acids being located mainly at the carboxy terminus.

B. BIOCHEMICAL ACTIVITY

The three human *ras* proteins have been purified using *E. coli* expression systems,[66-68] and if purified without denaturants they contain approximately 1 mol of bound GDP per mole protein.[69,70] The binding constants for GDP and GTP have been widely reported as being 10^{-8} M although one group reports much tighter binding at 2×10^{-11} M.[71]

The rate of exchange of the bound GDP molecule with exogenous nucleotide is very slow in Mg^{2+} containing buffers,[70] with a half-life of 60 min. However, we have found that in low Mg^{2+} (0.1 μM), the half-life of the ras.GDP complex in the presence of excess GTP is less than 1 min. On the face of it this looks analogous to the effect one might expect for *ras in vivo*, i.e., normal ras is in the resting state with GDP bound and exchanges very slowly even in the presence of the high intracellular concentrations of GTP (1 mM). After activation by a detector molecule (mimicked *in vitro* by removal of Mg^{2+}), exchange is speeded up and ras is converted into the GTP form. This is the activated state of the G protein and it would then be expected to interact with an effector molecule.

Consistent with a regulatory G protein role, the *ras* proteins have an intrinsic GTPase activity.[66,68,72] This is expected to convert the activated ras.GTP complex to inactivate ras.GDP. The half-life *in vitro* for hydrolysis of ras.GTP is very long (50 min; Kcat = 0.014 min^{-1}), which is similar to EF-Tu and around two orders of magnitude slower than for G_o, G_s, or transducin.[73] The significance of this low GTPase activity is not known but

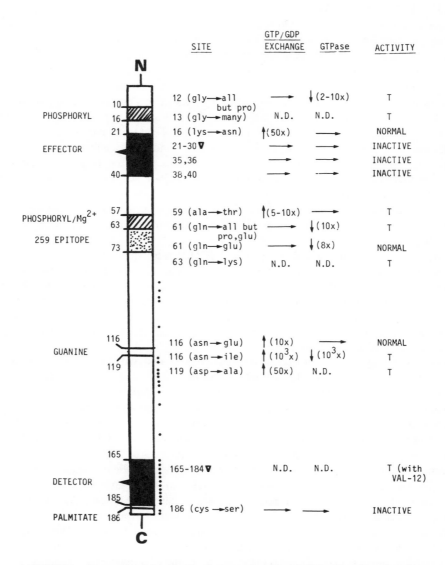

FIGURE 1. The proposed functional and structural regions of p21^{ras}. The regions have been defined using sequence comparison with EF-Tu protein and mutational analysis. Regions 10 to 16 and 57 to 63 should interact with the α/β phosphoryl groups of bound GTP/GDP. Region 57 to 63 might also be the Mg^{2+} binding site. The epitope to which the mouse monoclonal antibody 259 binds is also shown. The results of some mutational analyses are shown on the right. → = no change, ↑ = increase and ↓ = decrease relative to the normal protein. T = ability to transform NIH-3T3 cells without overexpression. The dots along the side of the protein represent positions of amino acid differences between any two *ras* proteins.

it seems incompatible with a regulatory G protein involved in a fast signaling process. In particular, the spontaneous GTP-GDP exchange rate is about equal to the measured GTPase rate leading to the conclusion that approximately 50% of normal p21 is constitutively in the active GTP form. This difficulty has now been resolved with the discovery of a cytoplasmic protein, GAP (GTPase activacting protein) which increases the GTPase activity of normal p21.[74] *In vivo,* i.e., in the presence of GAP, the steady state concentration of p21 in the active GTP form is predicted to be less than 2%.

C. MUTATIONAL ANALYSIS

The mutations found in human tumors at position 12, 13, or 61 have no effect on the binding constant for GTP or GDP.[68,72] However, it was first noticed that a Val-12 mutation in p21^{Ha-ras} caused a tenfold reduction in GTPase activity (see Figure 1).[66,67,72] Since the concentration of the active form of p21 (ras.GTP) in a cell is dependent on the relative rates of GTP/GDP exchange and GTP hydrolysis, it is likely that this mutant is always in the active form. This is somewhat analogous to the effect of cholera toxin on the α subunit of G$_s$.[75] In a study where all possible amino acid substitutions were made at codon 12, all except gly and pro were found to activate the protein.[76] In each case there was a general correlation between the reduction in GTPase activity and the transforming activity of the protein. A similar study at codon 61, however, came to an opposite conclusion.[77] All activating mutations produced a similar reduction in GTPase activity (8- to 10-fold), but the ability to transform NIH-3T3 cells differed greatly (over 1000-fold). Others have come to the same conclusion, namely that there is not always a correlation between the level of GTPase reduction (as measured *in vitro*) and the ability of the mutants to transform NIH-3T3 cells.[68,78]

The discovery of the cytoplasmic protein GAP has resolved these discrepencies.[74] It has been shown that the GTPase activities of Asp12, Val12,[74] and Lys61 (Calés and Hall, unpublished data) mutant p21s are not affected by GAP activity. Thus, although the differences in GTPase rates between Val12 or Asp12 mutants and normal are 10- and 2-fold, respectively, *in vitro,* in the presence of GAP, i.e., *in vivo,* the differences are around 300- and 60-fold. The GTPase rates of all mutants are therefore dramatically reduced compared to normal when measured *in vivo*.

Random mutagenesis of the normal *ras* gene has revealed two other sites that can activate the protein, namely 59 and 63.[79] Mutation and deletions have been introduced at other regions of the protein by site-directed mutagenesis (see Figure 1). Some alterations within the putative effector region (22 to 40) completely inactivate the biological activity of the protein without affecting GTP/GDP binding or GTPase activity.[61,62] Amino acid 116 and 119 are predicted to be in close contact with the guanine moiety of bound GDP/GTP.[58] Mutation of Asn116 to Ile resulted in no detectable binding of GTP to p21 (up to 10^{-5} M GTP) though surprisingly this mutation converted the normal gene to a transforming gene.[60] Similarly, the gene could be activated by conversion of Asp119 to Ala.[59] Alterations here affect the binding constants apparently by increasing the off rate for GDP (the rate-determining step for nucleotide exchange[70]). Again, this would affect the balance of ras.GDP and ras.GTP in cells leading to a high concentration of the active form. The same may be true for the threonine 59 mutation.[80,81] It is not yet known whether 116 and 119 mutations also affect the interaction of p21 with GAP. More recently, activation by alteration of amino acid 117 has been shown to occur in chemically induced rat liver carcinomas.[82] It is not known what the basis of this mutation is but it is likely to affect the binding constants for guanine nucleotides.

D. FUNCTION

Introduction of p21^{ras} protein into cells can have dramatic consequences. Microinjection of purified mutant proteins into NIH-3T3 cells causes changes in cell morphology around 15 h after injection.[52,68,83] This is followed by increased cell motility and cell division. Normal ras can induce the same transient changes but only at around 10× higher concentration. It is presumed that palmitoylation of the *E. coli*-produced ras occurs after microinjection. In a different kind of experiment, antibodies to p21^{ras} have been microinjected into quiescent rodent fibroblasts.[84] Subsequent addition of serum gave no stimulation of proliferation. These kinds of experiments strongly suggest that p21^{ras} is involved in controlling intracellular signaling elicited by a number of mitogens.

The signal generated by ras can, however, have consequences other than mitogenic stimulation. Microinjection of mutant p21^{ras} into xenopus oocytes for example stimulates cell maturation, mimicking the effects of either insulin or progesterone.[85] Interestingly, microinjection of antibodies to ras blocks the effect of insulin on oocyte maturation suggesting that the endogenous ras proteins of xenopus are involved in the normal insulin signaling pathway.[86] The antibodies have no effect on progesterone-stimulated maturation. Introduction of ras protein into the pheochromocytoma cell line PC12 induces terminal differentiation to nondividing neuronal cells, mimicking nerve growth factor.[87] It is assumed that a common biochemical signal is generated by ras in all these systems but that the cells "read" it differently. Some clues as to what this signal might be have recently been reported.

1. In Yeast

Deletion of both yeast *ras* genes is lethal and the introduction of a mutation into a yeast *ras* gene corresponding to an activating change in mammalian *ras* results in a defective response to nutritional deprivation.[88] There is now strong genetic and biochemical evidence that yeast RAS proteins are involved in the cyclic AMP effector pathway.[89] *In vitro* assays have been used to demonstrate that both purified yeast RAS and human p21^{H-ras} can activate adenylate cyclase activity in yeast membranes and that the active form of both normal and mutant ras in this assay is the GTP form.[90] Recently, an effector site mutation has been used to generate a poorly functioning yeast ras protein and the defect has been complemented by a mutation in a second protein.[91] Analysis has shown that this second site mutation is a single amino acid change in adenylate cyclase. Although not definitive proof, this strongly suggests a direct interaction between yeast ras and adenylate cyclase. There is some preliminary evidence that normal RAS function is dependent on the protein encoded by the CDC25 gene (a gene involved in cell division), and it has been postulated that CDC25 may function as the detector molecule for ras in yeast.[92]

2. In Mammalian Cells

A comparison of the putative effector regions of mammalian and yeast ras protein shows a high conservation of amino acids and mammalian ras is known to function in yeast. It is somewhat surprising, therefore, that ras apparently has no effect on the mammalian or amphibian adenylate cyclase systems.[85,93] For example, microinjection of mutant p21^{ras} into xenopus oocytes induces maturation without changes in cAMP levels, in contrast to progesterone which has the same biological effects but lowers cAMP levels. Accordingly, many groups have begun to look at other signaling systems, in particular, the phosphatidyl inositol (PI) pathway and activation of phospholipase C.

The introduction of mutant *ras* genes into NIH-3T3 results in increased levels of DAG and inositol phosphates.[94-96] This suggests an increase in PI turnover, and is similar to the situation found in polyoma virus transformed cells; though in that case PIP$_2$ levels were also higher.[44] To look at the affect of p21^{ras} on PI turnover in a more controllable and specific way, we have introduced the normal N-*ras* protooncogene under the control of an inducible promoter into NIH-3T3 cells.[96,97] We find that inducing the level of normal p21^{N-ras} up to 1% of the total membrane protein has no effect on the basal rate of PI turnover. However, if the cells are incubated in the presence of bombesin, a two- to threefold stimulation in inositol phosphate production was observed. Bombesin in the absence of increased p21^{N-ras} gives only a 20% stimulation. We conclude that p21^{N-ras} and activated bombesin receptor can synergize to stimulate phospholipase C, but at what level this synergy occurs is unknown. In agreement with others when we introduced an inducible mutant N-*ras* gene into NIH-3T3 cells, the basal rate of PI turnover was increased two- to fourfold when the transforming protein was switched on.[96,98]

Several groups have looked at second messenger changes in cells containing mutant Ha-*ras* genes and have come to somewhat different conclusions. One group reported that mutant Ha-*ras* inhibited PDGF-stimulated phospholipase C and A_2 activities,[99] while another group found a similar inhibition of PDGF-stimulated phospholipase C but an enhancement with bradykinin.[100] One of the major problems in working with transformed NIH-3T3 cells is that they produce their own growth factors including at least TGFα and PDGF-like molecules. Clearly a possible source of discrepancy between different groups could be the way in which the cells are handled and whether receptors are down regulated by autocrine production of factors. More recently another group has suggested that ras stimulates the production of diacylglycerol without production of inositol phospholipids[101] implying that other phospholipids such as phosphatidylcholine or phosphatidylethanolamine might be the targets for ras-stimulated breakdown. The consistent message from these results is that *ras* can affect phospholipid turnover; whether this is a primary effect or not is still unclear.

A mutational analysis which we have recently carried out[102] has suggested that a more direct approach might be possible to identify the target for regulation by ras. We have shown that mutations in the effector region (aa 30 to 40) of an otherwise normal p21 block the ability of GAP to increase its GTPase activity. This strongly suggests that GAP binds at the effector region of ras and therefore it may be the target for ras. If this is the case then it is to be expected that purification of GAP and sequence analysis will reveal the type of signaling process regulated by p21^{ras}.

E. *RAS* RELATED GENES

Using *ras* genes as probes, five other *ras*-related genes have been cloned from cDNA libraries; 3 *rho* genes (*rho*12, *rho*6, *rho*9),[103,104] R-*ras*,[105] and *ral*.[106] Figure 2 shows the amino acid sequence of Ha-*ras* compared to *rho*12, R-*ras,* and *ral.* It can be seen that the *rho* genes are more diverged from *ras* than R-*ras* or *ral.* Rho12 has 35% homology to human Ha-*ras,* and the homology is clustered and is especially high around the guanine nucleotide binding pocket. *rho*6 and *rho*12 are encoded in distinct genes and have 85% homology to each other with the highest divergence located in the carboxyterminal region. R-*ras* and *ral* have also been sequenced and are more closely related to Ha-*ras* (see Figure 2), with 55 and 50% homology, respectively. Again the homology is clustered around the guanine nucleotide domain. The amino acids corresponding to *ras* codons 12, 59, 61, 116, and 119 are conserved in all the genes, and it is likely that they each encode GTP/GDP binding proteins, though so far this has only been shown for R-*ras.*[105] Inspection of the putative effector region (20 to 40) of *ras* and the corresponding regions of the related genes shows that R-*ras* is very similar (15/20 amino acids identical). *ral* has 10/20 and *rho*12 6/20, identical amino acids. Whether these *ras*-related genes are regulatory G proteins involved in controlling cell growth will have to await further investigation.

V. CONCLUSION

The characterization of viral and cellular oncogenes has resulted in the identification of proteins involved in growth control. Although in the case of growth factor receptor oncogenes, some of these were already known, the involvement of tyrosine kinases and *ras* proteins in the signal transduction process was not predicted. Indeed we still do not known how they fit into the initial signaling events at the plasma membrane. There is a growing feeling that activation of phospholipase C and phospholipase A_2 are important early events in the action of many growth factor agonists and it seems likely that the tyrosine kinases and the ras proteins will somehow be involved in controlling these pathways.

```
Ha-ras    1                         MTEYKLVVVGAGGVGKSALTI
R-ras     1     MSSGAASGTGRGRPRGGGPGPGDPPPSETH ......G..........
ral       1                 MAANKPKGQNSLALH.VIM..S.........L
rho       1                     MAAIRK.L.I..D.AC..TC.L.


Ha-ras   22     QLIQNHFVDEYDPTIEDSYRKQVVIDGETCLLDILDTAGQEEYSAMR
R-ras    48     .F..SY..SD........T.ICSV..IPAR..........FG...
ral      33     .FMYDE..ED.E..KA.....K..L...EVQI.........D.A.I.
rho      24     VFSKDQ.PEV.V..VFEN.VADIEV..KQVE.ALW......D.DRL.


Ha-ras   69     DQYMRTGEGFLCVFAINNTKSFEDIHQYREQIKRVKDSDDVPMVLVG
R-ras    95     E....A.H...L.....DRQ..NEVGKLFT..L....R..F.V....
ral      80     .N.F.S........S.TEME..AATADF....L...EDEN..FL...
rho      71     PLSYPDTDVI.MC.S.DSPD.L.N.PELWTPEVK HFCPN..II...


Ha-ras  116     NKCDL           AA RTVESRQAQDLARSYGIPYIETSAKT
R-ras   142     ..A..           ESQ.Q.PRSE.SAFGA.HHVA.F.A...L
ral     127     ..S..           EDK.Q.SVEE.KNR.DQWNVN.V......
rho     117     ..K..RNDEHTRRELAKMKQEPVKPEEG RDMAN.IGAFG.M.C....


Ha-ras  149     RQGVEDAFYTLVREIRQHKLRKLNPPDESGPGC      MSCKCVLS
R-ras   176     .LN.DE..EQ...AV.KYQEQE.P.SPP.A.RKK      GGG.P...L
ral     161     .AN.DKV.FD.M....AR.MEDSKEKNGKKKRKSLAK RI RER.CIL
rho     164     KD..REV.EHAT.AAL.ARRG.KK              SG.LVL
```

FIGURE 2. A comparison of ras with the related proteins, R-ras, ral, and rho: single letter amino acid code has been used to compare the Ha-*ras* sequence with R-*ras, ral,* and *rho*12. Dots represent amino acids identical to Ha-*ras*.

It is still an enormous task to follow these signals through to the nucleus though the family of oncoproteins already contains cytoplasmic (mos, raf) and nuclear (myc, fos, jun) candidates for mediating this effect. Piecing together this jigsaw of mitogenic signaling is a daunting but exciting challenge. The future of targeted treatment of malignant disease surely depends upon it.

REFERENCES

1. **Hall, A.,** Oncogenes, in *Genetic Engineering,* Vol. 5, Rigby, P. W. J., Ed., Academic Press, New York, 1986, 61.
2. **Hunter, T.,** The proteins of oncogenes, *Sci. Am.,* 251(2), 60, 1986.
3. **Gilman, A. G.,** G proteins and dual control of adenylate cyclase, *Cell,* 36, 577, 1984.
4. **Watanabe, T., Umegaki, K., and Smith, W. L.,** Association of a solubilized prostaglandin E_2 receptor from renal medulla with a pertussis toxin-reactive guanine nucleotide regulatory protein, *J. Biol. Chem.,* 261, 13430, 1986.
5. **Berridge, M. J., Heslop, J. P., Irvine, R. F., and Brown, K. D.,** Inositol trisphosphate formation and calcium mobilization in Swiss 3T3 cells in response to platelet-derived growth factor, *Biochem. J.,* 222, 195, 1984.
6. **Macphee, C. M., Drummond, A. H., Otto, A. M., and Jiminez de Asna, L.,** Prostaglandin $F_{2\alpha}$ stimulates phosphatidylinositol turnover and increases the cellular content of 1,2 diacylglycerol in confluent resting Swiss 3T3 cells, *J. Cell. Physiol.,* 119, 35, 1984.
7. **Heslop, J. P., Blakeley, D. M., Brown, K. D., Irvine, R. F., and Berridge, M. J.,** Effects of bombesin and insulin on inositol (1,4,5) trisphosphate and inositol (1,3,4) trisphosphate formation in Swiss 3T3 cells, *Cell,* 47, 703, 1986.

8. **Nishizuka, Y.,** Turnover of inositol phospholipids and signal transduction, *Science,* 225, 1365, 1984.

9. **Litosch, I., Wallis, C., and Fain, J. N.,** 5-Hydroxytryptamine stimulates inositol phosphate production in cell-free systems from blowfly salivary glands, *J. Biol. Chem.,* 260, 5464, 1985.

10. **Ohta, H., Okajima, F., and Ui, M.,** Inhibition by islet-activating protein of a chemotactic peptide-induced early breakdown of inositol phospholipids and Ca^{2+} mobilization in guinea pig neutrophils, *J. Biol. Chem.,* 260, 15771, 1985.

11. **Ullrich, A., Coussens, L., Hayflick, J. S., Dull, T. J., Gray, A., Tam, A. W., Lee, J., Yarden, Y., Libermann, T. A., Schlessinger, J., Downward, J., Mayes, E. L. V., Whittle, N., Waterfield, M. D., and Seeberg, P. H.,** Human EGF receptor cDNA sequence and aberrant expression of the amplified gene in A431 epidermoid carcinoma cells, *Nature,* 309, 418, 1984.

12. **Nilsen, T. W., Maroney, P. A., Goodwin, R. G., Rothman, F. M., Crittenden, L. B., Raines, M. A., and Kung, H. J.,** c-*erb*B activation in ALV-induced erythroblastosis: novel RNA processing and promoter insertion result in expression of an amino-truncated EGF receptor, *Cell,* 41, 719, 1985.

13. **Riedel, H., Schlessinger, J., and Ullrich, A.,** A chimeric, ligand binding v-*erb*B/EGF receptor retains transforming potential, *Science,* 236, 197, 1987.

14. **Prywes, R., Livneh, E., Ullrich, A., and Schlessinger, J.,** Mutations in the cytoplasmic domain of the EGF receptor affect EGF binding and receptor internalization, *EMBO, J.,* 5, 2179, 1980.

15. **Downward, J., Waterfield, M. D., and Parker, P. J.,** Autophosphorylation and protein kinase C phosphorylation of the EGF receptor, *J. Biol. Chem.,* 260, 14538, 1985.

16. **Bertics, P. J. and Gill, G. N.,** Self-phosphorylation enhances the protein tyrosine kinase activity of EGF receptor, *J. Biol. Chem.,* 260, 14642, 1985.

17. **Hepler, J. R., Na Kahata, N., Lovenberg, T. W., Di Guiseppi, J., Herman, B., Earp, H. S., and Harden, T. K.,** Epidermal growth factor stimulates the rapid accumulation of inositol (1,4,5)-trisphosphate in A431 cells, *J. Biol. Chem.,* 262, 2951, 1987.

18. **Pepinsky, R. B. and Sinclair, L. K.,** Epidermal growth factor-dependent phosphorylation of lipocortin, *Nature,* 321, 81, 1986.

19. **Haigler, H. T., Schlaepfer, D. D., and Burgess, W. H.,** Characterization of Lipocortin I and an immunologically unrelated 33-KDa protein as epidermal growth factor receptor/kinase substrates and phospholipiase A$_2$ inhibitors, *J. Biol. Chem.,* 262, 6921, 1987.

20. **Schecter, A. L., Stern, D. F., Vaidyanathan, L., Decker, S. J., Drebin, J. A., Greene, M. I., and Weinberg, R. A.,** The *neu* oncogene: an *erb*-B related gene encoding a 185,000-Mr tumour antigen, *Nature,* 312, 513, 1984.

21. **Schechter, A. L., Hung, M. C., Vaidyanathan, L., Weinberg, R. A., Yang-Feng, T. L., Franke, U., Ullrich, A., and Coussens, L.,** The *neu* gene: an *erb*B-homologous gene distinct from and unlinked to the gene encoding the EGF receptor, *Science,* 229, 976, 1985.

22. **Bargman, C. I., Hung, M. C., and Weinberg, R. A.,** Multiple independent activations of the *neu* oncogene by a point mutation altering the transmembrane domain of p185, *Cell,* 45, 649, 1986.

23. **Yarden, Y. and Schlessinger, J.,** Self phosphorylation of the EGF receptor: evidence for a model of intermolecular allosteric activation, *Biochemistry,* 26, 1434, 1987.

24. **Scherr, C. J., Rettenmier, C. W., Sacca, R., Roussel, M. F., Look, A. T., and Stanlay, E. R.,** The c-*fms* proto-oncogene product is related to the receptor for mononuclear phagocyte growth factor, CSF-1, *Cell,* 41, 665, 1985.

25. **Wheeler, E. F., Rettenmeier, C. W., Look, A. T., and Scherr, C. J.,** The v-*fms* oncogene induces factor independence and tumorigenicity in CSF1 dependent macrophage cell line, *Nature,* 324, 377, 1986.

26. **Roussel, M. F., Dull, T. J., Rettenmeier, C. W., Ralph, P., Ullrich, A., and Sherr, C. J.,** Transforming potential of the c-*fms* proto-oncogene (CSF-1 receptor), *Nature,* 325, 549, 1987.

27. **Whetton, A. D., Monk, P. N., Consalvey, S. D., and Downes, D.,** The haemopoietic growth factors interleukin-3 and colony stimulating factor-2 stimulate proliferation but do not induce inositol lipid breakdown in murine bone-marrow-derived macrophages, *EMBO J.,* 5, 3287, 1986.

28. **Yarden, Y., Escobeda, J. A., Kuang, W. J., Yang-Feng, T. L., Daniel, T. O., Tremble, P. M., Chen, E. Y., Ando, M. E., Harkins, R. N., Francke, U., Fried, V. A., Ullrich, A., and Williams, L. T.,** Structure of the receptor for PDGF helps define a family of closely related growth factor receptors, *Nature,* 323, 226, 1986.

29. **Ullrich, A., Bell, J. R., Chen, E. Y., Herrera, R., Petrusselli, L. M., Dull, T. J., Gray, A., Coussens, L., Liao, Y. C., Tsubokana, M., Mason, A., Seeburg, P. H., Grunfeld, C., Rosen, O. M., and Ramachandran, J.,** Human insulin receptor and its relationship to the tyrosine kinase family of oncogenes, *Nature,* 313, 756, 1985.

30. **Wang, L. H., Lin, B., Jong, S. M. J., Dixon, D., Ellis, L., Roth, R. A., and Rutter, W. J.,** Activation of transforming potential of the human insulin receptor gene, *Proc. Natl. Acad. Sci. U.S.A.,* 84, 5725, 1987.

31. **Besmer, P., Murphy, J. E., George, P. C., Quin, F., Bergold, P. J., Lederman, L., Snyder, H. W., Brodeur, D., Zuckerman, E. L., and Hardy, W. D.,** A new acute feline retrovirus and relationship of its oncogene v-*kit* with the protein tyrosine kinase gene family, *Nature,* 320, 415, 1986.

32. **Young, D., Waitches, G., Birchmeier, C., Fasano, O. and Wigler, M.,** Isolation and characterization of a new cellular oncogene encoding a protein with multiple potential transmembrane domains, *Cell,* 45, 711, 1986.

33. **Takeya, T. and Hanafusa, H.,** DNA sequence of the viral and cellular *src* gene of chickens, *J. Virol.,* 44, 12, 1982.

34. **Parker, R. C., Varmus, H. E., and Bishop, J. M.,** Expression of v-*src* and chicken c-*src* in rat cells demonstrates qualitative differences between pp60$^{v\text{-}src}$ and pp60$^{c\text{-}src}$. *Cell,* 37, 131, 1984.

35. **Kmiecik, T. E. and Shalloway, D.,** Activation and suppression of pp60$^{c\text{-}src}$ transforming ability by mutation at its primary sites of tyrosine phosphorylation, *Cell,* 49, 65, 1987.

36. **Konopka, J. B. and Witte, O. N.,** Detection of c-*abl* tyrosine kinase activity *in vitro* permits direct comparison of normal and altered *abl* gene products, *Mol. Cell. Biol.,* 5, 3116, 1983.

37. **Jakobovits, E. B., Majors, J. E., and Varmus, H. E.,** Hormonal regulation of the Rous sarcoma virus *src* gene *via* a heterologous promoter defines a threshold dose for cellular transformation, *Cell,* 38, 757, 1984.

38. **Kamps, M. P., Buss, J. E., and Sefton, B. M.,** Rous sarcoma virus transforming protein lacking myristic acid phosphorylates known polypeptide substrates without inducing transformation, *Cell,* 45, 105, 1986.

39. **Calotty, G., Langier, D., Cross, F. R., Jovey, R., Hanafusa, T., and Hanafusa, H.,** The membrane-binding domain and myristylation of p60$^{v\text{-}src}$ are not essential for stimulation of cell proliferation, *J. Virol.,* 61, 1678, 1987.

40. **Huang, K. S., Wallner, B. P., Mattaleano, R. J., Tizard, R., Burne, C., Frey, A., Hession, C., McGray, P., Sinclair, L. K., Chow, E. P., Browning, J. L., Ramachandran, K. L., Tang, J., Smart, J. E., and Pepinsky, R. B.,** Two human 35kd inhibitors of phospholipase A$_2$ are related to substrates of pp60$^{v\text{-}src}$ and of the epidermal growth factor receptor kinase, *Cell,* 46, 191, 1986.

41. **Sugimoto, Y., Whitman, M., Cantley, L. C., and Erikson, R. L.,** Evidence that the rous sarcoma virus transforming gene product phosphorylates phosphatidylinositol and diacylglycerol, *Proc. Natl. Acad. Sci. U.S.A.,* 81, 2117, 1984.

42. **Macara, I. G., Marinetti, G. V., and Balduzzi, P. C.,** Transforming protein of avian sarcoma virus UR2 is associated with phosphatidylinositol kinase activity, *Proc. Natl. Acad. Sci. U.S.A.,* 81, 2728, 1984.

43. **Sugimoto, Y. and Erikson, R. L.,** Phosphatidylinositol kinase activities in normal and rous sarcoma virus-transformed cells, *Mol. Cell. Biol.,* 5, 3194, 1985.

44. **Whitman, M., Kaplan, D. R., Schaffhausen, B., Cantley, L., and Roberts, T.,** Association of phosphatidylinositol kinase activity with polyoma middle-T competent for transformation, *Nature,* 315, 239, 1985.

45. **Kaplan, D. R., Whitman, M., Schaffhausen, B., Pallas, D. C., White, M., Cantley, L., and Roberts, T. M.,** Common elements in growth factor stimulation and oncogenic transformation: 85kd phosphoprotein and phosphatidylinositol kinase activity, *Cell,* 50, 1021, 1987.

46. **Courtneidge, S. A. and Heber, A.,** A 81kd protein complexes with middle T antigen and pp60$^{c\text{-}src}$: a possible phosphatidylinositol kinase, *Cell,* 50, 1031, 1987.

47. **Hall, A.,** The ras gene family, in *Oxford Surveys of Eukaryotic Genes,* Vol. 1, Maclean, N., Ed., Oxford University Press, 1984, 111.

48. **Krontiris, T. G. and Cooper, G. M.,** Transforming activity of human tumor DNAs, *Proc. Natl. Acad. Sci. U.S.A.,* 78, 1181, 1981.

49. **Shih, C., Padhy, L. C., Murray, M., and Weinberg, R. A.,** Transforming genes of carcinomas and neuroblastomas introduced into mouse fibroblasts, *Nature,* 290, 261, 1981.

50. **Tabin, C. J., Bradley, S. M., Bargmann, C. I., Weinberg, R. A., Papageorge, A. G., Scolnick, E. M., Dhar, R., Lowy, D. R., and Chang, E. H.,** Mechanism of activation of a human oncogene, *Nature,* 300, 143, 1982.

51. **Bos, J. L., Toksoz, D., Marshall, C. J., Vries, M. V., Veeneman, G. H., Van der Eb, A. T., van Bloom J. H., Janssen, J. W. G., and Steenvoorden, A. C. M.,** Amino acid substitutions at codon 13 of the N-*ras* oncogene in human acute myeloid leukaemias, *Nature,* 315, 726, 1985.

52. **Stacy, D. N. and Kung, H. F.,** Transformation of NIH 3T3 cells by microinjection of Ha-*ras* p21 protein, *Nature,* 310, 508, 1984.

53. **Hall, A. and Marshall, C. J.,** unpublished data.

54. **Chang, E., Furth, M. E., Ellis, R. W., Scolnick, E. M., and Lowy, D. R.,** Tumorigenic transformation of mammalian cells induced by a normal human gene homologous to the oncogene of Harvey murine sarcoma virus, *Nature,* 297, 479, 1982.

55. **McKay, I. A., Marshall, C. J., Cales, C., and Hall, A.,** Transformation and stimulation of DNA synthesis in NIH-3T3 cells are a titratable function of normal p21$^{N\text{-}ras}$ expression, *EMBO J.,* 5, 2617, 1986.

56. **McGrath, J. P., Capon, D. J., Smith, D. M., Chen, E. V., Seeburg, P. H., Goeddel, D. V., and Levinson, A. D.,** Structure and organization of the human Ki-*ras* proto-oncogene and a related processed pseudogene, *Nature,* 304, 501, 1983.

57. **Hurley, J. B., Simon, M. I., Teplow, D. B., Robishaw, J. D., and Gilman, A. G.,** Homologies between signal transducing G proteins and *ras* gene products, *Science*, 226, 860, 1984.

58. **McCormick, F., Clark, B. F. C., LaCour, T. F. M., Kjeldgaard, M., Lauritsen, L. N., and Nyborg, J.,** A model for the tertiary structure of p21, the product of the *ras* oncogene, *Science*, 230, 78, 1986.

59. **Sigal, I. S., Gibbs, J. B., D'Alonzo, J. S., Temeles, G. L., Wolanski, B. S., Socher, S. H., and Scolnick, E. M.,** Mutant *ras*-encoded proteins with altered nucleotide binding exert dominant biological effects, *Proc. Natl. Acad. Sci. U.S.A.*, 83, 952, 1986.

60. **Walter, M., Clark, S. G., and Levinson, A. D.,** The oncogenic activation of human p21*ras* by a novel mechanism, *Science*, 233, 649, 1986.

61. **Willumsen, B. M., Papageorge, A. G., Kung, H. F., Bekesi, E., Robins, T., Johnsen, M., Vass, W. C., and Lowy, D. R.,** Mutational analysis of a *ras* catalytic domain, *Mol. Cell. Biol.*, 6, 2646, 1986.

62. **Sigal, I. S., Gibbs, J. B., D'Alonzo, J. S., and Scolnick, E. M.,** Identification of effector residues and a neutralizing epitope of Ha-*ras* encoded p21, *Proc. Natl. Acad. Sci. U.S.A.*, 83, 4725, 1986.

63. **Willumsen, B. M., Norris, K., Papageorge, A. G., Hubbert, N. L., and Lowy, D. R.,** Harvey murine sarcoma virus p21 *ras* protein: biological and biochemical significance of the cysteine residue nearest the carboxy terminus, *EMBO J.*, 3, 2581, 1984.

64. **Magee, A. I., Gutiervez, L., McKay, I. A., Marshall, C. J., and Hall, A.,** Dynamic fatty acylation of p21$^{N-ras.}$ *EMBO J.*, 6, 3353, 1987.

65. **Powers, S., Kataoka, T., Fasano, O., Goldfarb, M., Strathern, J., Broach, J., and Wigler, M.,** Genes in *S. cerevisiae* encoding proteins with domains homologous to the mammalian ras proteins, *Cell*, 36, 607, 1984.

66. **Gibbs, J. B., Sigal, I. S., Poe, M., and Scolnick, E. M.,** Intrinsic GTPase activity distinguishes normal and oncogenic *ras* p21 molecules, *Proc. Natl. Acad. Sci. U.S.A.*, 81, 5704, 1984.

67. **Sweet, R. W., Yokoyama, S., Kamata, T., Feramisco, J. R., Rosenberg, M., and Gross, M.,** The product of *ras* is a GTPase and the T24 oncogenic mutant is deficient in this activity, *Nature*, 311, 273, 1984.

68. **Trahey, M., Milley, R. J., Cole, G. E., Innis, M., Paterson, H., Marshall, C. J., Hall, A., and MacCormick, F.,** Biochemical and biological properties of the human N-*ras* p21 protein, *Mol. Cell. Biol.*, 7, 556, 1987.

69. **Poe, M., Scolnick, E. M., and Stein, R. B.,** Viral Harvey *ras* p21 expressed in *E. coli* purifies as a binary one to one complex with GDP, *J. Biol. Chem.*, 260, 3906, 1985.

70. **Hall, A. and Self, A. J.,** The effect of Mg^{2+} on the guanine nucleotide exchange rate of p21$^{N-ras.}$ *J. Biol. Chem.*, 261, 10963, 1986.

71. **Feuerstein, J., Goody, R. S., and Wittinghofer, A.,** Preparation and characterization of nucleotide-free and metal-ion free p21 "apoprotein", *J. Biol. Chem.*, 262, 8455, 1987.

72. **McGrath, J. P., Capon, D. J., Goeddel, D. V., and Levinson, A. D.,** Comparative biochemical properties of normal and activated human *ras* p21 protein, *Nature*, 310, 644, 1984.

73. **Higashijima, T., Ferguson, K. M., Smigel, M. D., and Gilman, A. G.,** The affect of GTP and Mg^{2+} on the GTPase activity and the fluorescent properties of G$_o$, *J. Biol. Chem.*, 262, 757, 1987.

74. **Trahey, M. and McCormick, F.,** A cytoplasmic protein stimulates normal N-*ras* p21 GTPase, but does not affect oncogenic mutants, *Science*, 238, 542, 1987.

75. **Birnbaumer, L., Swartz, T. L., Abramowitz, J., Mintz, P. W., and Iyengar, R.,** Transient and steady state kinetics of the interaction of guanyl nucleotides with the adenylate cyclase system from rat liver plasma membranes, *J. Biol. Chem.*, 255, 3542, 1980.

76. **Seeburg, P. H., Colby, W. W., Hayflick, J. S., Capon, D. J., Goeddel, D. V., and Levinson, A. D.,** Biological properties of human c-Ha-*ras*1 genes mutated at codon 12, *Nature*, 312, 71, 1984.

77. **Der, C. J., Finkel, T., and Cooper, G. M.,** Biological and biochemical properties of human ras^H genes mutated at codon 61, *Cell*, 144, 167, 1986.

78. **Lacal, J. C., Srivastava, S. K., Anderson, P. S., and Aaronson, S. A.,** Ras p21 proteins with high or low GTPase activity can efficiently transform NIH/3T3 cells, *Cell*, 44, 609, 1986.

79. **Fasano, O., Aldrich, T., Tamanoi, F., Taparowsky, E., Furth, M., and Wigler, M.,** Analysis of the transforming potential of the human H-*ras* gene by random mutagenesis, *Proc. Natl. Acad. Sci. U.S.A.*, 81, 4008, 1984.

80. **Hattori, S., Clarton, D. J., Satoh, T., Nakamura, S., Kaziro, Y,. Kawakita, M., and Shih, T. Y.,** Neutralizing monoclonal antibody against *ras* oncogene product p21 which impairs guanine nucleotide exchange, *Mol. Cell. Biol.*, 7, 1999, 1987.

81. **Lacal, J. C. and Aaronson, S. A.,** Activation of ras p21 transforming properties associated with an increase in the release rate of bound guanine nucleotide, *Mol. Cell. Biol.*, 6, 4214, 1986.

82. **Reynolds, S. H., Stowers, S. J., Patterson, R. M., Maronpot, R. R., Aaronson, S. A., and Anderson, M. W.,** Activated oncogenes in B6C3F1 mouse liver tumors: implication for risk assessment, *Science*, 237, 1309, 1987.

83. **Feramisco, J. R., Gross, M., Kamata, T., Rosenberg, M., and Sweet, R. W.,** Microinjection of the oncogene form of the human H-*ras* (T24) protein results in rapid proliferation of quiescent cells, *Cell,* 38, 109, 1984.

84. **Mulcahy, L. S., Smith, M. P., and Stacey, D. W.,** Requirement for *ras* proto-oncogene function during serum stimulated growth of NIH 3T3 cells, *Nature,* 313, 241, 1985.

85. **Birchmeier, C., Broek, D., and Wigler, M.,** Ras proteins can induce meiosis in xenopus oocytes, *Cell,* 43, 615, 1985.

86. **Korn, L. J., Siebel, C. W., McCormick, F., and Roth, R. A.,** *Ras* p21 as a potential mediator of insulin action in *Xenopus* oocytes, *Science,* 236, 840, 1987.

87. **Bar-Sagi, D. and Feramisco, J. R.,** Microinjection of the *ras* oncogene protein into PC12 cells induces morphological differentiation, *Cell,* 42, 841, 1985.

88. **Kataoka, T., Powers, S., McGill, C., Fasano, O., Strathern, J., Broach, J., and Wigler, M.,** Genetic analysis of yeast RAS1 and RAS2 genes, *Cell,* 37, 437, 1984.

89. **Toda, T., Uno, I., Ishikawa, T., Powers, S., Kataoka, T., Broek, D., Broach, J., Matsumoto, K., and Wigler, M.,** In yeast RAS proteins are controlling elements in the cyclic AMP pathway, *Cell,* 40, 27, 1985.

90. **Field, J., Broek, D., Kataoka, T., and Wigler, M.,** Guanine nucleotide activation of and competition between RAS proteins from *Saccharomyces cerevisiae, Mol. Cell. Biol.,* 7, 2128, 1987.

91. **Marshall, M. S., Gibbs, J. B., Scolnick, E. M., and Sigal, I. S.,** An adenylate cyclase from *Saccharomyces cerevisiae* that is stimulated by *RAS* proteins with effector mutations, *Mol. Cell. Biol.,* 8, 52, 1988.

92. **Broek, D., Toda, T., Michaeli, T., Levin, L., Birchmeier, C., Zoller, M., Powers, S., and Wigler, M.,** The *S. cerevisiae* CDC25 gene product regulates the RAS/adenylate cyclase pathway, *Cell,* 48, 789, 1987.

93. **Beckner, S. K., Hattori, S., and Shih, T. Y.,** The ras oncogene product is not a regulatory component of adenylate cyclase, *Nature,* 317, 71, 1985.

94. **Fleischman, L. F., Chahwala, S. B., and Cantley, L.,** Ras-transformed cells: altered levels of phosphatidylinositol 4,5 bisphosphate and catabolites, *Science,* 231, 407, 1986.

95. **Wolfman, A. and Macara, I. G.,** Elevated levels of diacylglycerol and decreased phorbol ester sensitivity in *ras*- transformed fibroblasts, *Nature,* 325, 359, 1987.

96. **Wakelam, M. J. O., Houslay, M. D., Davies, S. A., Marshall, C. J., and Hall, A.,** The role of N-*ras* p21 in the coupling of growth factor receptors to inositol phospholipid turn-over, *Biochem. Soc. Trans.,* 15, 45, 1987.

97. **Wakelam, M. J. O., Davies, S. A., Houslay, M. D., McKay, I., Marshall, C. J., and Hall, A.,** Normal p32^{N-ras} couples bombesin and other growth factor receptors to inositol phosphate production, *Nature,* 323, 173, 1986.

98. **Wakelam, M. J. O., McKay, I., Houslay, M. D., Hall, A., and Marshall, C. J.,** unpublished data.

99. **Benjamin, C. W., Tarpley, G. W., and Gorman, R. R.,** Loss of platelet derived growth factor-stimulated phospholipase activity in NIH-3T3 cells expressing the EJ-ras oncogene, *Proc. Natl. Acad. Sci. U.S.A.,* 84, 546, 1987.

100. **Parries, G., Hoebel, R., and Racker, I.,** Opposing effects of a *ras* oncogene or growth factor-stimulated phosphoinositide hydrolysis: densensitization to PDGF and enhanced sensitivity to bradykinin, *Proc. Natl. Acad. Sci. U.S.A.,* 84, 2648, 1987.

101. **Lacal, J. C., Moscat, J., and Aaronson, S. A.,** Novel source of 1,2 diacylglycerol elevated in cells transformed by Ha-*ras* oncogene, *Nature,* 330, 269, 1987.

102. **Calés, C., Hancock, J., Marshall, C. J., and Hall, A.,** unpublished data.

103. **Madaule, P. and Axel, R.,** A novel *ras*-related gene family, *Cell,* 41, 31, 1985.

104. **Yeramian, P., Chardin, P., Madaule, P., and Tavitian, A.,** Nucleotide sequence of human rho cDNA12, *Nucl. Acids Res.,* 15, 1869, 1987.

105. **Lowe, D. G., Capon, D. J., Delwart, E., Sakaguchi, A. Y., Naylor, S. L., and Goeddel, D. V.,** Structure of the human and mouse R-*ras* genes, *Cell,* 48, 13, 1987.

106. **Chardin, P. and Tavitian, A.,** The *ral* gene: a new *ras* related gene isolated by the use of a synthetic probe, *EMBO J.,* 2203, 1986.

Regulation of Phospholipase A$_2$ and C

Chapter 3

REGULATION OF PHOSPHOLIPASE A$_2$ BY G PROTEINS

Ronald M. Burch

TABLE OF CONTENTS

I.   Introduction 48

II.   G-Protein Regulation of Phospholipase A$_2$ 49
        A.   Pertussis Toxin-Sensitive G Protein 50
                1.   FRTL5 Cells 50
                2.   CPAE Cells 50
                3.   Rod Outer Segments 51
        B.   Pertussis Toxin-Insensitive G Protein................................ 51
        C.   Other Systems 52

III.   Phospholipase A$_2$ Coupled to Inhibitory G Proteins 53

IV.   Mechanism for G-Protein Regulation of Phospholipase A$_2$ 53

V.   Unanswered Questions 53

Acknowledgments 54

References 57

I. INTRODUCTION

Many hormones, growth factors, and certain neurotransmitters, when they bind to their receptors, stimulate the synthesis of a variety of biologically active products of arachidonic acid, the "eicosanoids", including prostaglandins, thromboxanes, leukotrienes, and hydroxyeicosatetraenoic acids.[1] Numerous studies have demonstrated that enhanced availability of arachidonic acid is the rate-limiting step in eicosanoid biosynthesis.[2] While there has been much debate over the relative roles of phospholipase A_2 or C in the release of arachidonic acid,[2,3,4] recent evidence from several cultured cell lines has clearly demonstrated that in those lines, stimulation of phospholipase A_2 by specific hormones liberates arachidonic acid for eicosanoid biosynthesis.[5-7]

Several mechanisms have been proposed to explain how phospholipase A_2 might be activated secondary to receptor occupation (Figure 1). Since many secretory phospholipases A_2 require calcium for maximum activity, a sequence has been proposed in which a receptor directly couples to phospholipase C to act on phosphatidylinositolbisphosphate to release inositol trisphosphate. Inositol trisphosphate then releases calcium from the endoplasmic reticulum to activate phospholipase A_2[8] (Figure 1, Scheme A). This sequence is doubtful as a general mechanism for activation of phospholipase A_2, since recent experiments in several cell lines have dissociated receptor activation of phospholipase C from phospholipase A_2.[5,6,9] In addition, in platelets, at least, there is evidence for a sequence in which a receptor is coupled directly to phospholipase A_2, and a product of phospholipase A_2 secondarily stimulates phospholipase C.[10] Also, the levels of calcium available after receptor activation (high nanomolar to low micromolar) are far below those usually found optimal for activation of secretory or total cellular phospholipase A_2 (several millimolar). In smooth muscle, however, increased calcium may be involved in direct activation of phospholipase A_2 since calcium channel antagonists may block receptor-stimulated arachidonic acid release[11] (Figure 1, Scheme B).

Other proposed mechanisms for activation of phospholipase A_2 rely on the observation that most nonlysosomal phospholipases A_2 are maximally active at alkaline pH. In fact, receptor activation in many cells does lead to increased intracellular pH due to the activity of an Na^+/H^+ counterport process, and in some,[12,13] but not all,[6] cells, inhibition of Na^+/H^+ counterport blocks receptor-mediated phospholipase A_2 activity (Figure 1, Scheme C).

Finally, phospholipase regulatory proteins have been described. The first is lipocortin, a phospholipase A_2 inhibitory protein.[14] Synthesis of this protein is stimulated by glucocorticoids. Inhibition of arachidonic acid release by glucocorticoids is an excellent marker for the involvement of phospholipase A_2 in this process since glucocorticoids are without effect on phospholipase C-catalyzed release of arachidonic acid.[6,15] This protein is thought to tonically inhibit phospholipase A_2. The current model for activation of a phospholipase A_2 regulated by lipocortin suggests that calcium-mobilizing hormones activate phospholipase A_2 indirectly by stimulating Ca^{2+}/phospholipid-dependent protein kinase (protein kinase C) to phosphorylate and inactivate lipocortin and activate phospholipase A_2[16] (Figure 1, Scheme D).

More recently, a phospholipase A_2-activating protein (PLAP) has been described.[17] For many years it has been recognized that protein synthesis inhibitors block receptor-mediated activation of phospholipase A_2. In several cell culture lines hormonal stimulation has been shown to lead to rapid synthesis of a PLAP which may directly activate phospholipase A_2[17] (Figure 1, Scheme E). Not considered in this review are other proteolytic mechanisms that may apply only to secretory phospholipases A_2.[18]

Recently, it has become clear that many cell surface receptors are coupled to a variety of effector enzymes by guanine nucleotide-binding regulatory proteins (G proteins) (this volume). In the present review, I will present evidence that G proteins also couple receptors to activation of phospholipase A_2.

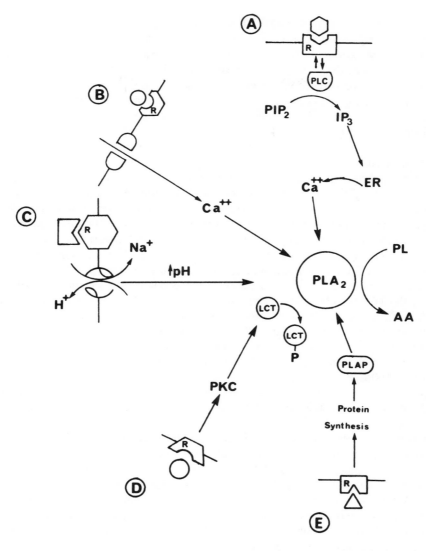

FIGURE 1. Mechanisms for regulation of phospholipase A_2. R, receptor; PLC, phospholipase C; PIP_2, phosphatidylinositolbisphosphate; IP_3, inositol trisphosphate; ER, endoplasmic reticulum; PLA_2, phospholipase A_2; PL, phospholipid; AA, arachidonic acid; LCT, lipocortin; PKC, protein kinase C; PLAP, phospholipase activating protein.

II. G-PROTEIN REGULATION OF PHOSPHOLIPASE A_2

The first indication that a G protein couples receptor activation to arachidonate release was the observation that pertussis toxin inhibits arachidonate release in response to the chemotactic peptide f-Met-Leu-Phe in neutrophils.[19] The only known mechanism of action of pertussis toxin is the ADP-ribosylation and inactivation of certain G proteins.[20] Similar observations were made by Ui and co-workers in neutrophils,[21,22] mast cells,[23,24] and Balb/c 3T3 cells.[25] In none of these studies was phospholipase A_2 activity examined directly. In several,[23,25] G-protein regulation of phospholipase A_2 was presented as the mechanism most compatible with the data since arachidonate release was pertussin toxin-sensitive but pharmacologically distinguishable from activation of phosphatidylinositol-specific phospholipase C.

A. PERTUSSIS TOXIN-SENSITIVE G PROTEIN

In studies of G-protein regulation of cellular activation, G proteins are conveniently divided into two groups, those which are pertussis toxin-sensitive, and those which are pertussis toxin-insensitive, based on whether their activation is blockable by pertussis toxin. After the description of G-protein regulation of adenylate cyclase,[26] G protein involvement in other transduction pathways has always been uncovered by the use of pertussis toxin as a probe.[27-29] When a transduction pathway utilizes a non-pertussis toxin-sensitive G protein, demonstration of G-protein involvement has always been difficult.

1. FRTL5 Cells

Direct evidence for a role for a G protein in the activation of phospholipase A_2 was obtained using the FRTL5 rat thyroid cell line.[5] In these cells, alpha$_1$-adrenergic receptor activation stimulates iodine release, organification of thyroglobulin, arachidonate release, phosphatidylinositol turnover, and DNA synthesis.[30,31] When FRTL5 cells were prelabeled with ^{3}H-arachidonic acid, norepinephrine caused a rapid release of free arachidonate[30] that could be blocked with the alpha$_1$-adrenergic antagonist prazosin, but not the alpha$_2$-adrenergic antagonist yohimbine, or the beta-adrenergic antagonist metoprolol.

To determine whether arachidonate release was coupled to receptor activation by a G protein, the effect of GTPγS, which causes persistent dissociation and activation of G proteins, was examined.[5] In cells transiently permeabilized by hypoosmotic treatment, GTPγS elicited release of ^{3}H-arachidonate. To further determine whether arachidonate release occurred by direct activation of phospholipase A_2, or required prior action of phosphatidylinositol-specific phospholipase C (see Figure 1, Scheme A), cells were prelabeled with ^{3}H-inositol, and phospholipase C activity was measured by the release of ^{3}H-inositol phosphate. GTPγS was found to stimulate phospholipase C. However, norepinephrine- and GTPγS-stimulated arachidonate release and phospholipase C activation were dissociated using neomycin. Neomycin inhibits phospholipase C activity by binding to the substrate, phosphatidylinositol.[32] After the cells had been treated with neomycin, neither norepinephrine or GTPγS was able to elicit inositol phosphate formation. However, in the presence of neomycin, both norepinephrine and GTPγS still stimulated arachidonate release. These results suggested that norepinephrine and GTPγS stimulate a phospholipase A_2 to cause arachidonate release. While the activity of GTPγS suggested that a G protein might be involved in the activation of phospholipase A_2, studies with pertussis toxin provided strong evidence. Pertussis toxin inhibited norepinephrine- and GTPγS-induced inositol phosphate formation[5] or increase in intracellular free calcium.[33]

Direct evidence for a G-protein-regulated phospholipase A_2 was obtained using membranes derived from FRTL5 cells and exogenous radiolabeled phosphatidylcholine substrates (Table 1). These experiments clearly demonstrated that norepinephrine and GTPγS can stimulate a pertussis toxin-sensitive, membrane-bound phospholipase A_2.

2. CPAE Cells

A pertussis toxin-sensitive G protein also mediates activation of phospholipase A_2 in response to leukotrienes in the CPAE bovine endothelial cell line.[34] In CPAE cells, leukotriene D_4 increased release of ^{3}H-arachidonate from cells which had been prelabeled with this fatty acid.[35] Using cell-free preparations, it was found that leukotriene D_4 increased the activity of a phosphatidylcholine-specific phospholipase A_2, but was without effect on phospholipase C, whether phosphatidylinositol, phosphatidylcholine, or phosphatidylethanolamine were used as substrates.[35] In these cells, as in FRTL5 cells, pertussis toxin blocked leukotriene-stimulated arachidonate release.[34] In CPAE cells the only pertussis toxin substrate apparent on electrophoretic gels after pertussis toxin-catalyzed ADP-ribosylation using ^{32}P-NAD had a molecular mass of about 41 kDa. When cells were incubated with increasing

TABLE 1
G-Protein Stimulation of Membrane-Bound
Phospholipase A₂ in FRTL5 Thyroid Cells

Treatment	Metabolite

2-[³H]Arachidonylphosphatidylcholine

	[³H]arachidonate
None	638 ± 76
GTPγS, 100 μM	1086 ± 260[a]
Norepinephrine, 1 μM	686 ± 54
Norepinephrine + GTPγS	2042 ± 320[b]

Phosphatidyl[³H]choline

	Lyso-[³H]phosphatidylcholine
None	428 ± 38
GTPγS	625 ± 120[a]
Norepinephrine	464 ± 72
Norepinephrine + GTPγS	979 ± 118[b]

Note: Cell membranes and assay were as in References 5 and 6.

[a] $p < 0.05$ compared to none.
[b] $p < 0.05$ compared to GTPγS alone.

concentrations of pertussis toxin, increasing amounts of ^{32}P-ADP were incorporated into the 41 kDa substrate. The dose-response for labeling of this substrate correlated exactly with pertussis toxin inhibition of leukotriene-stimulated arachidonate release.[34]

3. Retinal Rod Outer Segments

Perhaps the clearest evidence to date that a G protein is *directly* coupled to phospholipase A₂ was obtained in bovine rod outer segments.[36] The major G protein in rod outer segments is transducin.[37] In rod outer segments light activates its ''receptor'' rhodopsin, which activates transducin, leading to dissociation of its alpha subunit from its beta-gamma subunits. The dissociated, GTP-bound alpha subunit couples light to stimulation of cGMP-phosphodiesterase, leading to decreased concentration of cGMP and closure of the sodium channel.[37] The closing of the sodium channel causes hyperpolarization of the membrane and electrical activity which results in neurotransmission to the brain.

Using dark-adapted rod outer segments, light was found to activate phospholipase A₂.[36] G-protein mediation was suggested by the use of GTPγS, which mimicked light in the activation of phospholipase A₂. Strong evidence for transducin as the coupler of rhodopsin to phospholipase A₂ were the observations that both pertussis toxin and cholera toxin blocked light-induced activation of phospholipase A₂ as well as transducin-mediated, light-induced activation of cGMP-phosphodiesterase activity.[36]

Rod outer segments provide a powerful system for studying the role of G protein in the mediation of biological responses since transducin is readily removed from the membranes by hypotonic washing of dark-adapted rod outer segments. Using this property of rod outer segments, it was found that both light- and GTPγS-stimulated phospholipase A₂ activity was lost coincident with the removal of transducin. Most important, readdition of exogenous transducin to transducin-depleted rod outer segments partly restored light- and GTPγS-stimulated phospholipase A₂ activity.[36]

B. PERTUSSIS TOXIN-INSENSITIVE G PROTEIN

In the examples already described, the G protein implicated in the activation of phospholipase A_2 was pertussis toxin-sensitive. However, we have recently obtained evidence in Swiss 3T3 cells that pertussis toxin-insensitive G proteins can also couple receptors to activation of phospholipase A_2.[6] In Swiss 3T3 cells, bradykinin stimulates arachidonate release and phospholipase C-mediated inositol phosphate formation.[6] As in FRTL5 cells, arachidonate release could be dissociated from phosphatidylinositol-specific phospholipase C: in the 3T3 cells phorbol esters blocked bradykinin-induced inositol phosphate formation but enhanced bradykinin-stimulated arachidonate release. In contrast, bradykinin-stimulated arachidonate release was blocked by glucocorticoids and protein synthesis inhibitors, while inositol phosphate formation was unaffected. Involvement of phospholipase A_2 in arachidonate release was demonstrated in cells prelabeled with ^{3}H-choline, which was incorporated into phosphatidylcholine. When bradykinin was added, the phospholipase A_2 products lysophosphatidylcholine and glycerophosphocholine accumulated.

G-protein mediation of phospholipase A_2 activity was demonstrated by the ability of GTPγS to stimulate arachidonate release. In Swiss 3T3 cells, unlike FRTL5 cells, CPAE cells, or rod outer segments, the G protein responsible for activation of phospholipase A_2 was not pertussis toxin-sensitive. While pertussis toxin catalyzed ADP-ribosylation of a 40 kDa substrate in both intact Swiss 3T3 cells and membranes derived from them, it had no effect on either bradykinin- or GTPγS-stimulated arachidonate release.[6]

Since pertussis toxin did not offer the convenient probe for G-protein involvement in the activation of phospholipase A_2 and arachidonate release that it did in other cells, further evidence was sought. GDPβS is a guanine nucleotide analog which inhibits many G-protein-dependent reactions. Incubation of Swiss 3T3 cells with GDPβS inhibited subsequent bradykinin- or GTPγS-stimulated arachidonate release.[6]

C. OTHER SYSTEMS

In several other cell lines evidence is accumulating to suggest involvement of G proteins in the activation of phospholipase A_2. In MDCK canine kidney cells, alpha$_1$-adrenergic agonists stimulated arachidonate release and inositol phosphate formation.[38] As in the FRTL5 cells described above, the phospholipase C-catalyzed formation of inositol phosphates was pharmacologically distinguishable from arachidonate release. A pertussis toxin-insensitive G protein mediated arachidonate release since receptor binding was sensitive to guanine nucleotides. Further studies will be required to confirm phospholipase A_2 involvement. It would be useful to know whether arachidonate release is inhibited by glucocorticoids.

The involvement of a G protein in the activation of phospholipase A_2 by leukotrienes in CPAE cells was described above. In these cells also, it has been shown that tumor necrosis factor activates phospholipase A_2 in a pertussis toxin-sensitive fashion.[39] This cell line is of particular interest, since the leukotriene receptor and tumor necrosis factor receptor elicit arachidonate release by activating a pertussis toxin-sensitive phospholipase A_2, while bradykinin elicits arachidonate release by activating a phosphatidylcholine-specific phospholipase C[15] which is not pertussis toxin-sensitive.[34]

When human platelets were incubated with fluoroaluminate, which activates many G proteins,[40] inositol phosphates and arachidonate were released.[41] Several pharmacological manipulations distinguished the two processes, again suggesting parallel activation of two G proteins, one coupled to activation of phospholipase C, the other to arachidonate release (phospholipase A_2). Pertussis toxin blocked the ability of fluoroaluminate to stimulate arachidonate release.

In rabbit kidney proximal tubule cells, bradykinin and angiotensin II stimulated inositol phosphate formation and arachidonate release.[47] Arachidonate release was blocked by pertussis toxin. In these cells strong indirect evidence exists for G protein regulation of phospholipase A_2 since glucocorticoids blocked agonist-stimulated arachidonate release.[47]

III. PHOSPHOLIPASE A₂ COUPLED TO INHIBITORY G PROTEINS

Adenylate cyclase is under dual regulation by G proteins.[26] The alpha subunit of G_s activates adenylate cyclase, while the beta-gamma subunit of G_i binds G_s-alpha and leads to inhibition of adenylate cyclase. Indirect evidence suggests that phospholipase A_2 might also be under dual regulation by G proteins. Recall that GTPγS mimicked transducin in stimulating phospholipase A_2 activity in dark-adapted retinal rod outer segments. If, instead, rod outer segments were exposed to light to activate phospholipase A_2, then GTPγS added, the effect of GTPγS was to *inhibit* light-stimulated phospholipase A_2 activity.[36] While the data are indirect, they are consistent with GTPγS stimulating an inhibitory G protein to turn off the transducin-activated phospholipase A_2. Addition of GTPγS to transducin-depleted dark-adapted rod outer segments also uncovers the existence of the phospholipase A_2 inhibitory G protein, since in this condition, GTPγS also inhibited basal phospholipase A_2.[36]

In RAW264.7 macrophages, too, indirect evidence exists for dual regulation of phospholipase A_2 by stimulatory and inhibitory G proteins.[42] In these cells either cholera toxin or pertussis toxin stimulated phospholipase A_2 activity. If both toxins are added to the cells simultaneously, there is a synergistic stimulation of phospholipase A_2. Cholera toxin is known to activate certain G proteins, such as the stimulatory G_s coupled to adenylate cyclase, while pertussis toxin blocks activation of other G proteins, such as the inhibitory G_i coupled to adenylate cyclase. The resultant effect of either toxin on adenylate cyclase is stimulatory,[26] and when added together, stimulation may be much greater than the sum of their effects when added alone. Thus, while other models may be postulated for the effects of cholera toxin and pertussis toxin on phospholipase A_2 in RAW264.7 macrophages, a model which could account for the observations places phospholipase A_2 under dual regulation by stimulatory and inhibitory G proteins.

IV. MECHANISM FOR G-PROTEIN REGULATION OF PHOSPHOLIPASE A₂

The mechanism by which G proteins couple receptors to activation of phospholipase A_2 can be only speculative at present. G proteins appear to activate phospholipase C by reducing the calcium requirement of that enzyme.[43,44] Phospholipase A_2 is also a calcium-requiring enzyme. Evidence exists that phospholipase A_2 no longer requires calcium after being activated by a receptor.[45] Thus, reduction in the requirement for calcium may be the role played by a G protein in activation of phospholipase A_2.

Recently, it was reported that it is the beta-gamma subunits of transducin which activate phospholipase A_2 in the retina.[46] When added to dark-adapted, transducin-depleted rod outer segments, the beta-gamma subunits caused marked increase in phospholipase A_2 activity while addition of an equivalent amount of the alpha subunit caused only a slight increase. Addition of equimolar amounts of alpha with beta-gamma subunits inhibited beta-gamma-induced phospholipase A_2, presumably as a result of reassociation of the alpha-beta-gamma trimer. GTPγS, which prevents subunit association, abolished the inhibition of beta-gamma-stimulated phospholipase A_2 by the alpha subunit. Pertussis toxin pretreatment of the combined subunits, which causes irreversible association of the alpha-beta-gamma trimer, blocked phospholipase A_2 stimulation, even in the presence of GTPγS. Remaining to be defined is whether beta-gamma subunits directly activate phospholipase A_2 or whether they bind to and inactivate a phospholipase A_2 inhibitory alpha subunit.

V. UNANSWERED QUESTIONS

The concept of G-protein regulation of phospholipase A_2 is new. There are many questions related to this concept that remain to be answered before it can be generally accepted. Among the more important:

1.   How does a single class of receptor activate two separate transduction systems?
2.   Which G protein is involved?
3.   Do G proteins couple receptors directly to phospholipase A_2, or is the receptor coupled to a G protein which activates some process which ultimately results in increased phospholipase A_2 activity?
4.   What is the mechanism of the coupling?

The question of how, in a homogeneous cell line, a receptor for a single agonist activates two transduction systems is at present unclear. Based on the data presented here from several cell types, this may be a common phenomenon. In Figure 2 are presented several possible models. In Figure 2A, a single receptor has been represented, coupling to two transduction systems by two separate G proteins, one pertussis toxin-sensitive, the other, not. Such a situation in an actual cell has not been described and is difficult to envision. In Figure 2B, two separate receptors have been represented, one coupled to one G protein, the other to another G protein. The receptors could be subclasses of a single type, or, a single receptor type could couple to one of several transduction systems based on the relative abundance of the two transducers in the membrane. The work described that examined transducin activation of cGMP-phosphodiesterase and phospholipase A_2 suggests another possibility: activation of one transduction pathway by the G protein alpha subunit, and activation of another transduction pathway by the beta-gamma subunit. Such a mechanism is difficult to reconcile in the other systems described, in which one transduction system is pertussis toxin-sensitive while the other is not.

The G protein which couples receptors to activation of phospholipase A_2 appears to vary depending on the cell type. Some of the G proteins are G_i-like in that pertussis toxin blocks their activation. However, in Swiss 3T3 cells and MDCK cells, the G protein was not pertussis toxin-sensitive.

Perhaps the most important question to be answered is that of direct vs. indirect coupling. Much of the present discussion assumed direct receptor-G protein-phospholipase A_2 coupling. Certainly, in the studies of intact cells, direct coupling cannot be assumed. In Figure 3 are represented other potential coupling schemes in which receptors linked to G proteins could activate phospholipase A_2 indirectly. However, the studies which utilized transducin-depleted rod outer segments and readdition of purified transducin, and the studies of disrupted cell membranes (Table 1) do suggest direct coupling.

In conclusion, until purified hormone-sensitive phospholipase A_2 becomes available for reconstitution studies with purified G proteins, the directness of the coupling and the mechanism by which G-protein activation may activate phospholipase A_2 remains to be established. At this time the purification of a hormone-sensitive phospholipase A_2 appears to be the rate-limiting step.

ACKNOWLEDGMENTS

I would like to thank Drs. Julius Axelrod and Carole Jelsema for many stimulating discussions about phospholipase A_2 regulation.

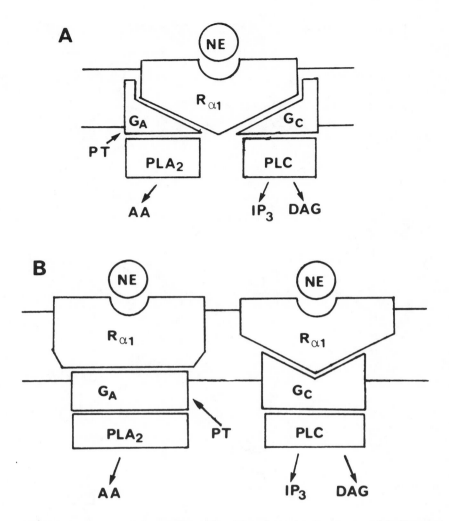

FIGURE 2. Mechanisms for coupling of the alpha$_1$-adrenergic receptor to parallel effector systems. G_A, G protein coupled to phospholipase A_2; G_c, G protein coupled to phospholipase C; PT, pertussis toxin; DAG, diacylglycerol; other abbreviations as in the legend to Figure 1.

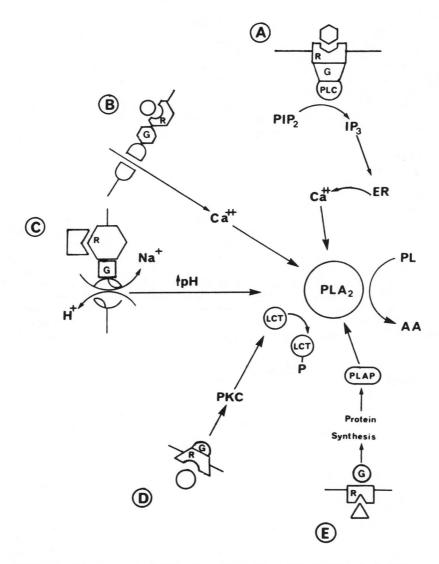

FIGURE 3.   Alternate mechanisms for G-protein regulation of phospholipase A_2. In Scheme D the G protein activates protein kinase C indirectly, through, for example, Scheme A.

REFERENCES

1. **Needleman, P., Turk, J., Jakschik, B. A., Morrison, A. R., and Lefkowith, J. B.,** Arachidonic acid metabolism, *Annu. Rev. Biochem.,* 55, 69, 1986.
2. **Irvine, R.,** How is the level of free arachidonic acid controlled in mammalian cells?, *Biochem. J.,* 204, 3, 1982.
3. **Majerus, P. W., Connolly, T. M., Deckmyn, H., Ross, T. S., Bross, T. E., Ishii, H., Bansal, V. S., and Wilson, D. B.,** The metabolism of phosphoinositide-derived messenger molecules, *Science,* 234, 1519, 1986.
4. **Lapetina, E. G.,** Regulation of arachidonic acid production: role of phospholipases C and A_2, *Trends Pharmacol. Sci.,* 3, 115, 1982.
5. **Burch, R. M., Luini, A., and Axelrod, J.,** Phospholipase A_2 and phospholipase C are activated by distinct GTP-binding proteins in response to α_1-adrenergic stimulation in FRTL5 thyroid cells, *Proc. Natl. Acad. Sci. U.S.A.,* 83, 7201, 1986.
6. **Burch, R. M. and Axelrod, J.,** Dissociation of bradykinin-induced prostaglandin formation from phosphatidylinositol turnover in Swiss 3T3 fibroblasts: evidence for G protein regulation of phospholipase A_2, *Proc. Natl. Acad. Sci. U.S.A.,* 84, 6374, 1987.
7. **Clark, M. A., Littlejohn, D., Conway, T. M., Mong, S., Steiner, S., and Crooke, S. T.,** Leukotriene D_4 treatment of bovine aortic endothelial cells and murine smooth muscle cells in culture results in an increase in phospholipase A_2 activity, *J. Biol. Chem.,* 261, 10713, 1986.
8. **Billah, M. M., Lapetina, E. G., and Cuatrecasas, P.,** Phospholipase A_2 activity specific for phosphatidic acid. A possible mechanism for the production of arachidonic acid in platelets, *J. Biol. Chem.,* 256, 5399, 1981.
9. **Slivka, S. R. and Insel, P. A.,** α_1-Adrenergic receptor-mediated phospholinositide hydrolysis and prostaglandin E_2 formation in Madin-Darby canine kidney cells. Possible parallel activation of phospholipase C and phospholipase A_2, *J. Biol. Chem.,* 262, 4200, 1987.
10. **Sweatt, J. D., Connolly, T. M., Cragoe, E. J., and Limbird, L. E.,** Evidence that Na^+/H^+ exchange regulates receptor-mediated phospholipase A_2 activation in human platelets, *J. Biol. Chem.,* 261, 8667, 1986.
11. **Satoh, H., Suzuki, J., and Satoh, S.,** Effects of calcium antagonist and calmodulin inhibitors on angiotensin II-induced prostaglandin production in the isolated dog renal artery, *Biochem. Biophys. Res. Commun.,* 126, 464, 1985.
12. **Sweatt, J. D., Blair, I., Cragoe, E. J., and Limbird, L. E.,** Inhibitors of Na^+/H^+ exchange block epinephrine- and ADP-induced stimulation of human platelet phospholipase C by blockade of arachidonic acid release at a prior step, *J. Biol. Chem.,* 261, 8660, 1986.
13. **Cuthbert, A. W. and Wong, P. Y. D.,** Calcium release in relation to permeability changes in toad bladder epithelium following antidiuretic hormone, *J. Physiol. London,* 241, 407, 1974.
14. **Hirata, F., Schiffmann, E., Venkatasubramanian, K., Salomon, D., and Axelrod, J.,** A phospholipase inhibitory protein in rabbit neutrophils induced by glucocorticoids, *Proc. Natl. Acad. Sci. U.S.A.,* 77, 2533, 1980.
15. **Clark, M. A., Bomalaski, J. S., Conway, T., Wartell, J., and Crooke, S. T.,** Differential effects of aspirin and dexamethasone on phospholipase A_2 and C activities and arachidonic acid release from endothelial cells in response to bradykinin and leukotriene D_4, *Prostaglandins,* 32, 703, 1986.
16. **Hirata, F.,** The regulation of lipomodulin, a phospholipase inhibitory protein in rabbit neutrophils by phosphorylation, *J. Biol. Chem.,* 256, 7730, 1981.
17. **Clark, M. A., Conway, T. M., Shorr, R. G. L., and Crooke, S. T.,** Identification and isolation of a mammalian protein which is antigenically and functionally related to the phospholipase A_2 stimulatory peptide mellitin, *J. Biol. Chem.,* 262, 4402, 1987.
18. **Van den Bosch, H.,** Intracellular phospholipases A_2, *Biochem. Biophys. Res. Commun.,* 604, 191, 1980.
19. **Bokoch, G. M. and Gilman, A. G.,** Inhibition of receptor-mediated release of arachidonic acid by pertussis toxin, *Cell,* 39, 301, 1984.
20. **Ui, M.,** Islet-activating protein, pertussis toxin, *Trends Pharmacol. Sci.,* 5, 277, 1984.
21. **Okajima, F. and Ui, M.,** ADP-ribosylation of the specific membrane protein by islet-activating protein, pertussis toxin, associated with inhibition of a chemotactic peptide-induced arachidonate release in neutrophils. A possible role of the toxin substrate in Ca^{2+}-mobilizing biosignaling, *J. Biol. Chem.,* 259, 13863, 1984.
22. **Ohta, H., Okajima, F., and Ui, M.,** Inhibition by islet-activating protein of a chemotactic peptide-induced early breakdown of inositol phospholipids and Ca^{2+} mobilization in guinea pig neutrophils, *J. Biol. Chem.,* 260, 15771, 1985.
23. **Nakamura, T. and Ui, M.,** Islet-activating protein, pertussis toxin, inhibits Ca^{2+}-independent and guanine nucleotide-dependent releases of histamine and arachidonic acid from rat mast cells, *FEBS Lett.,* 173, 414, 1984.

24. **Nakamura, T. and Ui, M.,** Simultaneous inhibitions of inositol phospholipid breakdown, arachidonic acid release, and histamine secretion in mast cells by islet-activating protein, pertussis toxin. A possible involvement of the toxin-specific substrate in the Ca^{2+}-mobilizing receptor-mediated biosignaling system, *J. Biol. Chem.*, 260, 3584, 1985.

25. **Murayama, T. and Ui, M.,** Receptor-mediated inhibition of adenylate cyclase and stimulation of arachidonic acid release in 3T3 fibroblasts. Selective susceptibility to islet-activating protein, pertussis toxin, *J. Biol. Chem.*, 260, 7226, 1985.

26. **Gilman, A. G.,** G proteins and the dual regulation of adenylate cyclase, *Cell,* 36, 577, 1984.

27. **Cockcroft, S.,** Role of guanine nucleotide binding protein in the activation of polyphosphoinositide phosphodiesterase, *Trends Biochem. Sci.,* 12, 75, 1987.

28. **Yatani, A., Codina, J., Brown, A. M., and Birnbaumer, L.,** Direct activation of mammalian atrial muscarinic potassium channel by the GTP regulatory protein G_k, *Science,* 235, 207, 1987.

29. **Holz, G. G., IV, Rane, S. G., and Dunlap, K.,** GTP-binding proteins mediate transmitter inhibition of voltage-dependent calcium channels, *Nature,* 319, 470, 1986.

30. **Burch, R. M., Luini, A., Mais, D., Corda, D., Vanderhoek, A., Kohn, L., and Axelrod, J.,** α_1-Adrenergic stimulation of arachidonic acid release and metabolism in a rat thyroid cell line. Mediation of cell replication by prostaglandin E_2, *J. Biol. Chem.,* 261, 11236, 1986.

31. **Kohn, L. D., Aloj, S. M., Tombaccini, D., Rotella, C. M., Toccafondi, R., Marcocci, C., Corda, D., and Grollman, E. F.,** The thyrotropin receptor, in *Biochemical Actions of Hormones,* Vol. 12, Litwack, G., Ed., Academic Press, Orlando, FL, 1985, 457.

32. **Schacht, J.,** Purification of polyphosphoinositides by chromatography on immobilized neomycin, *J. Lipid Res.,* 19, 1063, 1978.

33. **Corda, D. and Kohn, L. D.,** Role of pertussis toxin sensitive G protein in the alpha$_1$ adrenergic receptor but not thyrotropin receptor mediating the activity of membrane phospholipases and iodide fluxes in FRTL5 thyroid cells, *Biochem. Biophys. Res. Commun.,* 141, 1000, 1986.

34. **Clark, M. A., Conway, T. M., Bennett, C. F., Crooke, S. T., and Stadel, J. M.,** Islet-activating protein inhibits leukotriene D_4- and leukotriene C_4- but not bradykinin- or calcium ionophore-induced prostacyclin synthesis in bovine endothelial cells, *Proc. Natl. Acad. Sci. U.S.A.,* 83, 9320, 1986.

35. **Clark, M. A., Littlejohn, D., Conway, T. M., Mong, S., Steiner, S., and Crooke, S. T.,** Leukotriene D_4 treatment of bovine aortic endothelial cells and murine smooth muscle cells in culture results in an increase in phospholipase A_2 activity, *J. Biol. Chem.,* 261, 10713, 1986.

36. **Jelsema, C. L.,** Light activation of phospholipase A_2 in rod outer segments of bovine retina and its modulation by GTP-binding proteins, *J. Biol. Chem.,* 262, 163, 1987.

37. **Stryer, L.,** The molecules of visual excitation, *Sci. Am.,* 257, 42, 1987.

38. **Slivka, S. R. and Insel, P. A.,** α_1-Adrenergic receptor-mediated phosphoinositide hydrolysis and prostaglandin E_2 formation in Madin-Darby canine kidney cells. Possible parallel activation of phospholipase C and phospholipase A_2, *J. Biol. Chem.,* 262, 4200, 1987.

39. **Clark, M. A., Chen, M. J., and Bomalaski, J. S.,** Tumor necrosis factor induces phospholipase A_2 activation and the synthesis of a phospholipase activity protein in endothelial cells, *Fed. Proc.,* 46, 1946, 1987.

40. **Sternweis, P. and Gilman, A. G.,** Aluminum: a requirement for activation of the regulatory component of adenylate cyclase by fluoride, *Proc. Natl. Acad. Sci. U.S.A.,* 79, 4888, 1982.

41. **Fuse, I. and Tai, H.-H.,** Stimulation of arachidonate release and inositol-1,4,5-triphosphate formation are mediated by distinct G-proteins in human platelets, *Biochem. Biophys. Res. Commun.,* 146, 659, 1987.

42. **Burch, R. M., Jelsema, C., and Axelrod, J.,** Cholera toxin and pertussis toxin stimulate prostaglandin E_2 synthesis in a murine macrophaze cell line, *J. Pharmacol. Exp. Ther.,* 245, 765, 1988.

43. **Smith, C. D., Cox, C. C., and Snyderman, R.,** Receptor-coupled activation of phosphoinositide-specific phospholipase C by an N protein, *Science,* 232, 97, 1986.

44. **Bradford, P. G. and Rubin, R. P.,** Guanine nucleotide regulation of phospholipase C activity in permeabilized rabbit neutrophils, *Biochem. J.,* 239, 97, 1986.

45. **Bormann, B. J., Huang, C.-K., Mackin, W. M., and Becker, E. L.,** Receptor-mediated activation of phospholipase A_2 in rabbit neutrophil plasma membrane, *Proc. Natl. Acad. Sci. U.S.A.,* 18, 767, 1984.

46. **Jelsema, C. L. and Axelrod, J.,** Stimulation of phospholipase A_2 activity in bovine rod outer segments by the β,γ subunits of transducin and its inhibition by the α subunit, *Proc. Natl. Acad. Sci. U.S.A.,* 84, 3623, 1987.

47. **Welsh, C., Dubyak, G., and Douglas, J. G.,** Relationship between phospholipase C activity and prostaglandin E_2 and cyclic adenosine monophosphate production in rabbit tubular epithelial cells, *J. Clin. Invest.,* 81, 710, 1988.

Chapter 4

THE INVOLVEMENT OF A GTP-BINDING REGULATORY PROTEIN IN MUSCARINIC RECEPTOR-STIMULATED INOSITOL PHOSPHOLIPID METABOLISM

Lawrence A. Quilliam and Joan Heller Brown

TABLE OF CONTENTS

I.   Introduction 60

II.   Involvement of a G Protein in Phospholipase C Activation 60
        A.   G Proteins and Adenylate Cyclase 60
        B.   Astrocytoma Cells and Muscarinic Receptor-Stimulated PI
                Hydrolysis 60
        C.   G Proteins Couple to Ca^{2+}-Mobilizing Receptors 61

III.   Regulation of G_{PLC} 63
        A.   Regulation of G_{PLC} Function by Bacterial Toxins 63
        B.   Regulation of G_{PLC} Function by Phosphorylation 65

IV.   Identification of G_{PLC} 67
        A.   Identification of G_{PLC} by Reconstitution 67
        B.   Identification of G_{PLC} by Covalent Modification 69

V.   Conclusion 70

Acknowledgments 70

Abbreviations 70

References 70

I. INTRODUCTION

Many cellular responses to hormones, neurotransmitters, and mitogens are mediated by the hydrolysis of phosphoinositides. This results in the generation of the second messenger molecules, inositol 1,4,5-trisophosphate and diacylglycerol, which promote the release of Ca^{2+} from intracellular stores and activate the Ca^{2+}- and phospholipid-dependent protein kinase (protein kinase C), respectively. The details of this process have been the topic of several recent reviews.[1-4] The hydrolysis of phosphoinositides is catalyzed by a PI-specific phosphodiesterase, phospholipase C which, at physiological Ca^{2+} concentrations preferentially breaks down the polyphosphoinositides. Until recently, the regulation of phospholipase C activity was believed to be due to subtle changes in its lipid environment, thus altering its structural conformation and/or presenting the phosphoinositides as better substrates.[5,6] Recent studies have demonstrated however that a GTP-binding protein, analogous to that regulating adenylate cyclase, is involved in the coupling of receptors to phospholipase C.

II. INVOLVEMENT OF A G PROTEIN IN PHOSPHOLIPASE C ACTIVATION

A. G PROTEINS AND ADENYLATE CYCLASE

The involvement of GTP binding (G) proteins in the regulation of second messenger production was initially discovered for adenylate cyclase. Much of the knowledge obtained from investigating its regulation has been used to study the involvement of G proteins in PI hydrolysis.

Stimulatory and inhibitory receptors which regulate the generation of cyclic AMP by adenylate cyclase are coupled to the catalytic unit by the oligomeric GTP binding proteins, G_s and G_i, respectively. These proteins consist of unique α subunits with apparent molecular weights of between 40 and 52 kDa plus common β and γ subunits of approximately 35 and 8 kDa, respectively. The α subunits bind guanine nucleotides and contain a GTPase activity. G_s is inactive when the guanine nucleotide binding site of the α subunit is occupied by GDP. Upon the binding of stimulatory agonist, receptors promote the exchange of GDP for GTP bound to the α_s subunit of the holoprotein and induce subunit dissociation. This nucleotide exchange reaction is associated with a decrease in receptor-agonist affinity. The free, activated α_s GTP is responsible for stimulation of the catalytic moiety. Hydrolysis of GTP to GDP by the intrinsic GTPase results in inactivation of G_s and reformation of the heterotrimer that can then undergo further cycles. G_i undergoes a simiar cycle when it is activated by inhibitory agonist-receptor coupling. The major mechanism of inhibition of adenylate cyclase is believed to be through the dissociation of G_i to yield a $\beta\gamma$ complex that can interact with and activate α_s. The inhibitory α subunit can also directly inhibit cyclic AMP production since it is effective in membranes from the cyc$^-$ variant of S49 lymphoma cells (which do not possess α_s).

The use of two bacterial toxins that transfer the ADP moiety of NAD^+ onto the α subunits of G proteins has greatly aided in the isolation and characterization of the G proteins that interact with adenylate cyclase. Cholera toxin was found to ADP-ribosylate α_s resulting in inhibition of the intrinsic GTPase activity and persistent activation of adenylate cyclase. The pertussis toxin (islet activating protein) of *Bordetella pertussis* ADP-ribosylates the α subunit of G_i, locking it in the $\alpha\beta\gamma$ form and preventing hormonal inhibition of adenylate cyclase. Further details on the above processes can be found in several excellent reviews.[7-9]

B. ASTROCYTOMA CELLS AND MUSCARINIC RECEPTOR-STIMULATED PI HYDROLYSIS

Muscarinic cholinergic receptors are known to inhibit adenylate cyclase through G_i. In the 1321N1 human astrocytoma cell line muscarinic agonists can inhibit cyclic AMP accumulation, however Harden and coworkers demonstrated that this response was blocked in the presence of cyclic nucleotide phosphodiesterase inhibitors.[10] Also, treatment of these cells with pertussis toxin, which catalyzes the ADP-ribosylation of $G_i\alpha$, preventing hormonal inhibition of adenylate cyclase, did not affect carbachol-mediated inhibition of cyclic AMP accumulation.[11] It was, therefore, concluded that muscarinic receptors do not directly inhibit adenylate cyclase or couple to G_i in astrocytoma cells. The decrease in cyclic AMP accumulation in these cells has since been shown to be mediated by the activation of a Ca^{2+}-calmodulin-dependent phosphodiesterase.[12]

The ability of muscarinic receptors to stimulate PI metabolism was initially demonstrated by Hokin and Hokin[13] and the relationship of phosphoinositides and Ca^{2+} metabolism to muscarinic receptor activation was further suggested by Michell and colleagues.[14] The possibility that muscarinic receptor-mediated stimulation of the Ca^{2+}-calmodulin dependent phosphodiesterase in astrocytoma cells was secondary to PI hydrolysis was therefore considered. Indeed, in 1321N1 cells carbachol stimulated PI hydrolysis with the concomitant release of inositol phosphates.[15] Time courses of inositol phosphate accumulation revealed that $InsP_3$ and $InsP_2$ were formed earlier than $InsP$, which accumulated linearly with time in the presence of the InsP phosphomonoesterase inhibitory Li^+. This suggested that the polyphosphoinositides were the initial substrate for phospholipase C resulting in the production of $InsP_3$ which was subsequently dephosphorylated to inositol. More detailed analysis of the inositol phosphates demonstrated that both Ins $1,4,5$-P_3 and Ins $1,3,4$-P_3 were formed rapidly following addition of carbachol.[16,17] $InsP_4$ also appeared extremely rapidly suggesting that the Ins $1,4,5$-P_3 formed from the initial breakdown of phosphatidylinositol 4,5-bisphosphate is rapidly shunted through $InsP_4$ to Ins $1,3,4$-P_3. Carbachol also stimulates the efflux of $^{45}Ca^{2+}$ from 1321N1 cells[15] and recent studies with fura-2-loaded 1321N1 cells have shown that carbachol induces a rapid increase in the intracellular Ca^{2+} concentration, independently of extracellular Ca^{2+}.[18]

C. G PROTEINS COUPLE TO Ca^{2+}-MOBILIZING RECEPTORS

Radioligand binding studies as early as 1974 demonstrated that the binding of agonists to receptors whose actions were believed to be mediated via Ca^{2+} mobilization, rather than through adenylate cyclase, were regulated by guanine nucleotides.[19] This suggested that these receptors might be coupled to a GTP-binding protein(s) analogous to adenylate cyclase (see Reference 20). Although the lack of effect of pertussis toxin on cyclic AMP accumulation in 1321N1 cells suggested that muscarinic receptors in these cells were not coupled to G_i, the binding of muscarinic agonists to receptors in washed membranes prepared from 1321N1 cells was regulated by guanine nucleotides.[21] High- and low-affinity agonist binding states were observed, as seen with adenylate cyclase regulating receptors. In addition, Evans et al. demonstrated that the relative capacity of several cholinergic agonists to form a GTP-sensitive, high affinity binding state was directly related to their efficacy to stimulate PI hydrolysis and Ca^{2+} efflux (Figure 1).[22] These results strongly implicated the involvement of a G protein in the coupling of muscarinic receptors to phospholipase C in 1321N1 cells.

About this time Haslam and Davidson demonstrated that GTPγS could stimulate the accumulation of diacylglycerol in permeabilized platelets, suggesting that a G protein might interact with and stimulate phospholipase C.[23] Direct evidence for stimulation of PI hydrolysis by guanine nucleotides was first reported in 1985 by Litosch et al.[24] who demonstrated that GTP was necessary for the coupling of the 5-hydroxytryptamine receptor of blowfly salivary gland membranes to PI hydrolysis and by Cockcroft and Gomperts[25] who demonstrated that GTPγS stimulated the hydrolysis of polyphosphoinositides in neutrophil membranes. The putative G protein(s) coupling receptors to phospholipase C will be referred to as G_{PLC} in this chapter.

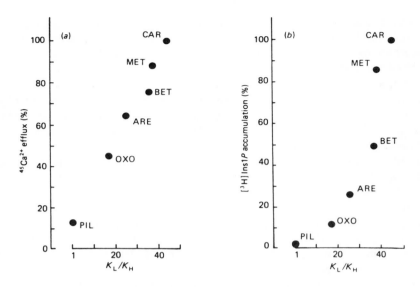

FIGURE 1. Relationship between the ratio of dissociation constants for low- and high-affinity binding states and efficacy of muscarinic agonists for stimulation of Ca^{2+} efflux and PI breakdown. The ratio of dissociation constants for low and high affinity binding states (K_L/K_H) for the indicated agonists is plotted vs. the efficacy of these agonists (relative to carbachol) for stimulation of Ca^{2+} efflux (A) or PI breakdown (B). Abbreviations: CAR, carbachol; MET, methacholine; BET, bethanecol; ARE, arecoline; OXO, oxotremorine; PIL, pilocarpine. (Reprinted with permission from Evans, T., Hepler, J. R., Masters, S. B., Brown, J. H., and Harden, T. K., *Biochem. J.*, 232, 751, 1985. The Biochemical Society, London.)

Studies on washed membranes prepared from 1321N1 cells prelabeled with [^{3}H] inositol also demonstrated that GTPγS could stimulate PI hydrolysis.[26] Carbachol was ineffective at causing inositol phosphate formation except in the presence of GTPγS, suggesting that a G protein was obligatory for coupling of the muscarinic receptor to phospholipase C. In addition, GTPγS was necessary for the stimulation of inositol phosphate accumulation induced by histamine and bradykinin. Similar results were obtained by Hepler and Harden who also reported that GTP could support carbachol-stimulated Ins 1,4,5-P$_3$ formation in 1321N1 cells.[27]

Fluoride, probably in the form of AlF_4^-,[28] stimulates G-protein subunit dissociation and activation in a manner analogous to nonhydrolyzable GTP analogs.[28,29] Bigay et al. have suggested that this effect is due to the interaction of AlF_4^- with GDP in the nucleotide binding site where it mimics the γ phosphate of GTP.[30] The dose dependency for NaF-stimulated PI hydrolysis in 1321N1 cell membranes was found to be similar to that reported for stimulation of adenylate cyclase activity and was potentiated by micromolar concentrations of $AlCl_3$.[27,31] In more purified 1321N1 membranes we have found that GTPγS, carbachol plus GTP, and AlF_4^- preferentially stimulate polyphosphoinositide hydrolysis (Figure 2) as occurs in response to hormone *in vivo*. The stimulatory effect of AlF_4^- on PI hydrolysis (Figure 3, open bars), unlike its effect on cyclic AMP production (not shown), was also observed in intact cells. This difference in sensitivity of adenylate cyclase vs. phospholipase C to AlF_4^- supports the notion that the G protein responsible for coupling receptors to phospholipase C is functionally distinct from G_s and G_i.

Muscarinic stimulation of PI hydrolysis has also been studied in permeabilized cells and membranes prepared from several other cell types.[32-37] Inositol phosphate formation induced by carbachol was shown to be potentiated by guanine nucleotides in isolated chick heart cells,[32] exocrine pancreas,[33] pancreatic islet cells,[34] and smooth muscle cells.[36] In brain

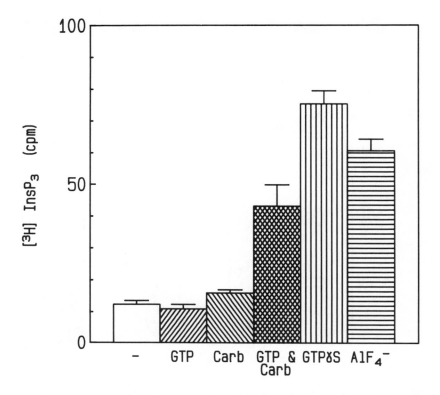

FIGURE 2.   Effects of agents on [³H] InsP₃ formation from 1321N1 astrocytoma cell membranes. Washed membranes, prepared from cells prelabeled with [³H] inositol, were incubated with guanine nucleotides (100 μM) and/or carbachol (0.5 mM) or with NaF (10 mM) plus A1Cl₃ (10 μM). [³H] InsP₃ was separated by anion exchange chromatography. Values are mean ± SD, n = 3.

membranes, guanine nucleotides[37,38] and carbachol[37] stimulated PI hydrolysis, but potentiation of hormonal stimulation by guanine nucleotides was not observed,[37] probably due to the presence of sufficient endogenous GTP in the preparation. Guanine nucleotide dependence of PI hydrolysis has been documented for a wide variety of other agonists (see Reference 39).

III. REGULATION OF G_{PLC}

A. REGULATION OF G_{PLC} FUNCTION BY BACTERIAL TOXINS

In a number of cell types, particularly those derived from bone marrow, pretreatment with pertussis toxin has been shown to inhibit agonist-induced stimulation of PI hydrolysis (see Table 1). This suggested that G_i or a related pertussis toxin substrate may couple receptors to phospholipase C in these cells. In many other cells however (Table 1) pertussis toxin has been reported to have no effect on agonist-induced PI breakdown.

Muscarinic receptor-stimulated PI hydrolysis is unaffected by pertussis toxin in a number of cell types including astrocytoma cells,[52] chick heart cells,[52] pancreas,[33,34] brain,[53] pituitary,[50] and aortic smooth muscle.[36] In many of these tissues pertussis toxin has been shown to effectively block muscarinic inhibition of adenylate cyclase. For example, in chick heart cells pertussis toxin (100 ng/ml) can completely block muscarinic inhibition of cyclic AMP production but has no effect on carbachol-stimulated inositol phosphate accumulation at concentrations up to tenfold higher. In 1321N1 cells, pretreatment with pertussis toxin results

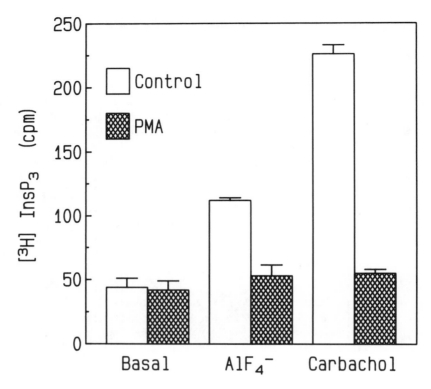

FIGURE 3. Effects of carbachol and AlF_4^- on PI hydrolysis in 1321N1 cells and attenuation of responses by phorbol ester. Cells prelabeled with [^{3}H] inositol were incubated for 15 min in the presence of absence of PMA (1 μM) prior to the addition of A1C1$_3$ (10 μM)/NaF (20 mM) or carbachol (0.5 mM) for a further 15 min. Values are mean ± SD, n = 3.

TABLE 1
Sensitivity of Cells to Inhibition of PI Hydrolysis
by Pertussis Toxin

Inhibited by pertussis toxin	Not inhibited by pertussis toxin
Neutrophil[40]	*Pituitary tumor cells[50,51]
HL-60 cells[41]	*Astrocytoma cells[52]
Mast cell[42]	*Heart cells[32,53]
Leukocytes[43]	*Brain[54]
Macrophages[44]	Mammary tumor cells[55]
Renal mesangial cells[45]	*Pancreatic cells[33,34]
Liver[46]	Liver[46,56]
Vascular smooth muscle[47]	*Vascular smooth muscle[36]
Neuronal cells[48]	Neuronal cells[57]
Hamster fibroblasts[49]	3T3 fibroblasts[58]
	Brown adipocytes[59]

Note: Asterisk indicates systems in which muscarinic responses have been examined.

in greater than 95% ADP-ribosylation of G proteins of approximately 41 kDa (based on inhibition of the subsequent incorporation of [^{32}P] ADP into membrane proteins) without inhibiting PI hydrolysis.[52]

Interestingly, pertussis toxin has been reported to stimulate PI hydrolysis in some cell types.[55,58] These observations might be explained by a recent study on human platelets where pertussis toxin, at high concentrations, was found to stimulate inositol phosphate formation independently of ADP-ribosylation.[60]

The lack of effect of pertussis toxin on PI hydrolysis in so many cell types suggests that different G proteins, at least one of which is insensitive to pertussis toxin, couple different receptors to phospholipase C. On the other hand, it can be argued that G_{PLC} may be a substrate for pertussis toxin even though toxin treatment does not perturb PI breakdown. For example, in hepatocytes where G_i is in excess of angiotensin II receptors, no attenuation of agonist-induced cyclic AMP production was observed until more than 80% of G_i had been ribosylated.[61] Therefore, in studies where 100% G-protein ribosylation is not achieved, there may be sufficient active G_{PLC} to maximally stimulate phospholipase C. Also $G_i\alpha$ has a higher affinity for the $\beta\gamma$ complex and is a better substrate for pertussis toxin than G_o, a GTP-binding protein originally isolated from bovine brain.[62] Thus, in cells where the concentration of the common pool of $\beta\gamma$ complex is limiting, G_i is more likely to exist as a heterotrimer and to be ADP-ribosylated than G_o or possibly other pertussis toxin substrates which might couple receptors to phospholipase C. Further data regarding the pertussis toxin sensitivity of the G_{PLC} that couples muscarinic receptors to phospholipase C are discussed in Section IV.A.

Cholera toxin pretreatment did not affect PI hydrolysis in a variety of cell types tested including astrocytoma,[31,63] GH_3 pituitary,[51] and pancreatic acinar cells.[33] Interestingly, cholera toxin has recently been reported to block inositol phosphate accumulation in some other cell types,[47,55,64,65] In the human embryonic pituitary tumor cell line, Flow 9000, the toxin was shown to attenuate muscarinic agonist- and cholecystokinin-stimulated PI hydrolysis.[65] In membranes prepared from these cells, cholera toxin pretreatment did not attenuate the stimulatory effects of GTPγS on PI hydrolysis but significantly reduced the GTPγS-potentiated cholecystokinin response. Thus, cholera toxin may affect receptor-G protein coupling in these cells. Cholera toxin has also been shown to affect phosphatidylinositol monophosphate levels in A431 and rat liver cells while having no significant effect on the formation of inositol phosphate.[66,67] Since this effect was not due to elevation of the intracellular cyclic AMP concentration, these observations raise the possibility that a PI kinase is also regulated by a G protein.

B. REGULATION OF G_{PLC} FUNCTION BY PHOSPHORYLATION

In addition to the diacylglycerol formed by PI hydrolysis *in vivo,* protein kinase C is activated by several classes of tumor promoter including the phorbol esters.[4] Preincubation of 1321N1 cells with PMA had no effect on the basal level of PI hydrolysis but completely blocked accumulation of $InsP_3$, $InsP_2$, and $InsP$ in response to carbachol.[66] The concomitant blockade of $^{45}Ca^{2+}$ efflux from PMA treated cells and of the increase in intracellular Ca^{2+}-concentration measured by fura-2 fluorescence suggested a causal relationship between $InsP_3$ formation and Ca^{2+} mobilization.[18,68] The K_D of the muscarinic receptor for [^{3}H] methyl scopolamine was unchanged by PMA treatment of 1321N1 cells while the IC_{50} for carbachol was only modestly increased.[67] In addition, PMA treatment did not alter the guanine nucleotide-induced shift in agonist binding to muscarinic receptors in washed membranes.[26] Similarly, Vicentini et al. reported that PMA treatment attenuated carbachol-stimulated inositol phosphate formation in PC12 cells, but had no effect on binding competition curves for muscarinic agonists.[69] The stimulation of inositol phosphate formation in response to bradykinin and histamine in 1321N1 cells was also blocked by PMA suggesting that protein kinase C acts at a common site in the pathway, distal to agonist-receptor coupling (Figure 4). This notion was further supported by the observation that fluoride-stimulated PI hydrolysis in intact cells was also blocked by PMA pretreatment (Figure 3).

GTPγS-stimulated PI hydrolysis in astrocytoma cell membranes was found to be inhibited by PMA pretreatment.[26] Similarly, the addition of purified protein kinase C to 1321N1 cell

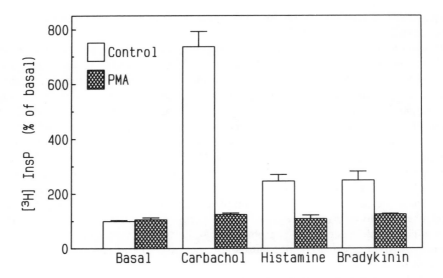

FIGURE 4. Phorbol ester blocks hormone-dependent [³H] InsP formation in 1321N1 cells. Cells prelabeled with [³H] inositol were treated for 15 min with PMA (1 μ*M*). Cells were then incubated with carbachol (100 μ*M*), histamine (100 μ*M*) or bradykinin (1 μ*M*) for 15 min and were assayed for [³H] InsP formation. Values are mean ± SEM, n = 3 to 8.

membranes resulted in a blockade of inositol phosphate formation in response to GTPγS. PMA was shown to block GTPγS-stimulated PI breakdown by 50 to 75% in membrane preparations, whereas stimulation by millimolar Ca^{2+} was only minimally reduced. This suggested that protein kinase C affected the ability of G_{PLC} to couple to phospholipase C rather than having a direct effect on the activity of the phospholipase. Guanine nucleotides have been shown to shift the dose-response of Ca^{2+}-stimulated inositol phosphate formation to the left, suggesting that the activated G protein might serve to sensitize phospholipase C to the action of Ca^{2+} (which acts at the level of the phospholipase and/or its substrate).[70,71] Thus, it is possible that the small decrease in Ca^{2+}-stimulated PI hydrolysis induced by PMA in 1321N1 cells is due to attenuation of the G_{PLC}:phospholipase C interaction. Smith et al. similarly concluded that PMA affected the coupling of G_{PLC} to phospholipase C in polymorphonuclear leukocytes; although PMA attentuated f-Met-Leu-Phe-stimulated PI hydrolysis, the phorbol ester had no effect on the rate of the chemoattractant-stimulated binding of GTPγS to membranes nor on Ca^{2+}-stimulated inositol phosphate formation.[72] A schematic model of phospholipase C regulation by G_{PLC} and protein kinase C in 1321N1 cells is shown in Figure 5.

Phorbol ester-induced attenuation of PI hydrolysis is a common phenomenon, reported for many cell types.[4] The degree of attenuation of the PI response by PMA in different cell types does however vary. In 1321N1 cells, for example, muscarinic agonist-induced PI hydrolysis is completely inhibited by PMA whereas in NCB-20 neurohybridoma and PC-12 pheochromocytoma cells inhibition of the muscarinic response is only partial.[68,69,73] Further, in some cells no effect of phorbol esters on agonist-stimulated PI hydrolysis has been observed.[48,74] The observation that PMA attenuates PI hydrolysis in pertussis toxin-sensitive, e.g., leukocytes[72] and -insensitive, e.g., astrocytoma cells[67] suggests that different coupling proteins have a common site for regulation by protein kinase C or that the site of phosphorylation is distal to the G protein. Phosphorylation of phospholipase C in guinea pig uterus has recently been shown to be stimulated by phosbol esters *in vivo*, thus, phospholipase C rather than G_{PLC} might be the site of regulation of the PI pathway by protein kinase C.[75] Another possible mechanism of decreasing inositol phosphate levels is by stim-

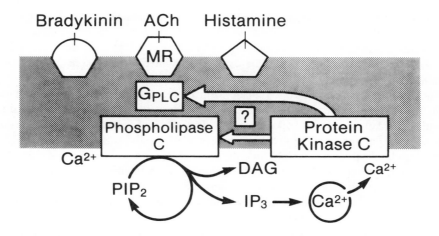

FIGURE 5. Schematic representation of phospholipase C regulation in 1321N1 cells. Abbreviations: ACH acetylcholine; DAG, diacyglycerol; MR, muscarinic cholinergic receptor; PIP₂, phosphatidylinositol 4,5-bisphosphate.

ulating their degradation. Although phorbol esters have been shown to stimulate platelet InsP₃ 5′-phosphomonoesterase activity,[76,77] PMA had no effect on [³H] Ins 1,4,5-P₃ hydrolysis in 1321N1 cells.[26]

The inhibitory effect of protein kinase C represents a potential mechanism of negative feedback regulation of PI hydrolysis by diacylglycerol production. In 1321N1 cells, however, no short term desensitization of PI hydrolysis in response to carbachol occurs; InsP accumulation is linear for at least 80 min in the presence of Li⁺.[15] Protein kinase C can be downregulated by prolonged exposure of cells to phorbol esters.[78] Down regulation of protein kinase C by chronic treatment of 1321N1 cells with PMA prevented the acute inhibitory effect of the phorbol ester on PI breakdown but basal- and carbachol-stimulated inositol phosphate levels were not elevated.[31,78] These observations suggest a lack of feedback inhibition by protein kinase C in 1321N1 cells under normal conditions of stimulation. In contrast, long-term treatment of Swiss 3T3 fibroblasts with PMA resulted in an increase in basal- and bombesin-stimulated InsP levels.[79] Thus, in this cell line there appears to be tonic inhibition of PI hydrolysis by protein kinase C. However, it is possible that these differences between cell types are due to the expression of different isozymes of protein kinase C that show varying resistance to down regulation by phorbol esters rather than to the expression of unique G_{PLC}s.[80]

IV. IDENTIFICATION OF G_{PLC}

A. IDENTIFICATION OF G_{PLC} BY RECONSTITUTION

As discussed above, it seems likely that there is more than one G protein capable of coupling receptors to phospholipase C. One approach taken to address this possibility is the purification of receptors, G proteins and phospholipase C followed by reconstitution. Agonist binding to muscarinic receptors has been shown to be guanine nucleotide dependent. This effect is lost when the muscarinic receptor is purified but can be reconstituted by the addition of purified G_i, G_o, or G_N, a protein purified from neutrophil membranes.[81-83] This suggests that muscarinic receptors are able to couple to a number of different, related G proteins. Ui and colleagues have demonstrated the ability of purified G_i and G_o to couple f-Met-Leu-Phe receptors to phospholipase C in HL-60 cells after inactivation of native G proteins with pertussis toxin.[41] G_i, G_o, and a 40-kDa pertussis toxin substrate recently purified from rat brain membrane was also reported to stimulate a partially purified phospholipase C from platelet membranes.[84] These experiments support the data that a G protein, having homology with G_i, interacts with phospholipase C in these cells. However, since all three G proteins tested were equally effective in stimulating PI hydrolysis, whereas all three are unlikely to serve as G_{PLC} *in vivo,* caution must be exercised in interpreting reconstitution studies as reflecting G-protein function.

Another, perhaps more physiological, form of reconstitution is the use of molecular cloning followed by expression of receptors and/or G proteins in cells normally lacking these molecules. Ashkenazi and co-workers recently demonstrated that an M_2 muscarinic receptor clone, when expressed in Chinese hamster ovary (CHO) cells, stimulated PI hydrolysis and inhibited cyclic AMP formation.[85] Thus, the same receptor could couple to either adenylate cyclase or phospholipase C. Surprisingly, pertussis toxin inhibited muscarinic stimulation of PI hydrolysis, although a considerably higher concentration of the toxin was required to completely block the stimulation of phospholipase C than adenylate cyclase activity. The dose response curve for ribosylation was also quite shallow, suggesting the involvement of two different G proteins. Since only one ADP-ribosylated protein band was detected on gels following toxin treatment, the two G proteins must be of very similar apparent molecular weight, e.g., two forms of G_i may couple the muscarinic receptor to the two enzymes. The high level of recombinant receptor expression (1.45×10^6 receptors per cell) could also allow the muscarinic receptor to couple to phospholipase C through a G protein that it would not interact with at the lower receptor levels found on most cells *in vivo.*

In another form of reconstitution, *Xenopus* oocytes were injected with rat brain messenger RNA.[86] This resulted in the expression of muscarinic receptors which stimulated PI hydrolysis and a transient chloride current in the host cells. The increase in chloride conductance was mimicked by injecting cells with Ins $1,4,5$-P_3, suggesting that the muscarinic response was secondary to PI hydrolysis. Since muscarinic stimulation of the current transients was blocked by pertussis toxin, the authors suggested that PI breakdown is mediated by G_i or G_o. Thus, in the oocyte and the transfected CHO cells the muscarinic receptor-stimulated PI response expressed is sensitive to pertussis toxin, although muscarinic-stimulated PI hydrolysis has not been shown to be pertussis toxin-sensitive in any normally coupled systems.

At least four distinct muscarinic receptor subtypes have been cloned from porcine, rat, and human sources.[87-89] It may, therefore, be that certain muscarinic receptor subtypes interact with pertussis toxin-sensitive G proteins *in vivo* to open K^+ channels and to inhibit adenylate cyclase, while other receptor subtypes interact with unidentified pertussis toxin-insensitive G protein(s) that activate phospholipase C. However, since a single muscarinic receptor subtype can be shown to couple to three different G proteins,[81-83] and at least three different purified G proteins can couple to phospholipase C[41,84] (we are aware of additional unpublished reports), it is possible that these interactions are not faithful. Instead, G proteins may interact with several receptors and effectors with differing efficiency. Thus, following reconstitution with a limited number of receptors, G proteins may couple them in an abnormal way causing an unexpected biochemical response. The overexpression of one or more components of the signal transduction apparatus or co-expression of these components in the wrong cell might also result in a novel cellular response.

The *ras* proteins are a group of GTP-binding proteins that have molecular weights of approximately 21 kDa. These proteins share considerable sequence homology with the α subunits of the known GTP binding signal transducing proteins. They also have a similar GTPase activity although the activated or oncogenic forms of p21 characteristically have a much lower enzyme activity. Unlike G_s, G_i, and transducin, however, p21^{ras} proteins do not appear to interact with $\beta\gamma$ subunits.[90] Some cells that have been transfected with *ras* genes have been reported to have a higher level of PI hydrolysis than control cells[91,92] or to have an increased PI response to growth factors[93] and muscarinic agonists.[94] In the latter report, transfected cells had a basal level of inositol phosphates similar to control cells but had a fivefold greater response to carbachol. It is not clear if these observations reflect direct coupling of receptors to phospholipase C by p21^{ras} or some other change in cell metabolism that occurs as a consequence of expression of *ras* gene products. Since the carbachol-induced increase in inositol phosphate formation in *ras*-transfected fibroblasts was attenuated by treatment with pertussis toxin, which does not ADP-ribosylate p21^{ras}, it is possible that the *ras* gene products modulate the function of another G protein.

Recent experiments by Aaronson and colleagues have overcome the potential problem of side effects due to long-term maintainance of *ras*-transfected cells in culture. They found that increases in diacylglycerol and inositol phosphate levels were detectable within minutes of microinjecting *Xenopus* oocytes with p21^{r-ras} but not p21^{c-ras}.[95] Thus, p21 can influence the hydrolysis of inositol (and possibly other) phospholipids without inducing the synthesis of other proteins. Further experiments are required to determine whether these *ras* gene products couple directly to phospholipase C or merely interfere with other G protein functions.

B. IDENTIFICATION OF G$_{PLC}$ BY COVALENT MODIFICATION

Due to the apparent lack of ribosylation of G$_{PLC}$ by cholera and pertussis toxins, it is proving difficult to identify the novel G protein(s) that couple muscarinic receptors to phospholipase C. A number of putative G proteins have been suggested by GTP-binding methods, but it is difficult to identify their function.[96-98] A covalent modification that could prove useful in detecting G$_{PLC}$ is protein phosphorylation. Our data suggest that protein kinase C modulates PI hydrolysis at the level of G protein:phospholipase C coupling[26,31] and there are precedents for G-protein phosphorylation by protein kinase C, eukaryotic initiation factor 2 and G$_i$ being substrates for this enzyme.[99,100] Following overnight labeling with [^{32}P] orthophosphate, 1321N1 cells were challenged with PMA or carbachol. Antibodies raised against various G proteins were then used to immunoprecipitate these proteins from detergent-solubilized cell extracts and the proteins separated by SDS gel electrophoresis. These experiments demonstrated the presence of G$_S\alpha$, G$_i\alpha$, p21^{ras}, and Gα_{25} (G$_p$) in 1321N1 cells.[101] A 40-kDa protein, G$_i\alpha_2$,[102] was also present. No increase in the phosphorylation state of any of the aforementioned immunoprecipitated G proteins was observed in response to PMA or carbachol. It is not clear from these results whether G$_{PLC}$ lacks sequence homology with other G proteins (and therefore fails to be precipitated by any of the five G-protein antibodies tested) or whether it does not get phosphorylated. Interestingly, p21^{ras} was found to be phosphorylated even in the absence of stimulation.

The identification of a toxin that invariably affects the function of the G protein regulating phospholipase C would be of great experimental value. Botulinum neurotoxin types C and D were recently shown to ADP-ribosylate a protein of 20 to 26 kDa in several cell types.[103,104] Further, the cation and nucleotide requirements for ribosylation in 1321N1 cell membranes suggested that the type-D toxin substrate was a G protein.[31] The similarity in molecular weights of p21^{ras} and the botulinum toxin type-D substrate suggested that these proteins might be identical. However, the mobility of the ribosylated toxin substrate was different from that of [^{35}S] methionine-labeled p21^{ras} from the same cells. Furthermore, the ribosylated substrate was not immunoprecipitated by a monoclonal antibody (Y13-259) directed against p21^{ras}.[105] To determine whether the protein ribosylated by botulinum toxin type D regulates phospholiase C, membranes from [^{3}H] inositol-labeled 1321N1 cells were treated with the toxin prior to measuring inositol phosphate formation. No correlation of ADP-ribosylation with PI hydrolysis was found.[31] Thus, a covalent modification to aid in the isolation and characterization of the pertussis toxin insensitive G$_{PLC}$ remains elusive. Also antibodies raised against protein sequences common to known G proteins have so far failed to recognize a potential G$_{PLC}$ molecule. The recent copurification of GTP binding activity along with receptors that regulate PI hydrolysis and with phospholipase C suggests that G$_{PLC}$(s) might soon be isolated.[84,106,107]

V. CONCLUSION

A great deal has been learned about the G-protein regulation of PI hydrolysis over the past 3 years. With the isolation of receptors, G proteins and phospholipases C[108,109] and their genes we can expect further major advances to be made in our understanding of receptor-effector coupling. It is clear, however, that caution must be exercised in the interpretation of data obtained from reconstitution and genetic studies.

ACKNOWLEDGMENTS

This work was supported by University of California Cancer Research Coordinating Committee funds and by National Institutes of Health Grant GM36927.

ABBREVIATIONS

G protein — GTP-binding regulatory protein
G_{PLC} — Unidentified G protein(s) stimulating phospholipase C
G_s, G_i — Stimulatory and inhibitory G proteins of adenylate cyclase, respectively
G_o — A 40-kDa G protein purified from bovine brain
$G\alpha_{25}$ — A G protein purified from platelets and placenta (G_p)
GTPγS — Guanosine 5'-(3-O-thiotriphosphate)
InsP — Inositol monophosphate
$InsP_2$ — Inositol bisphosphate
$InsP_3$ — Inositol trisphosphate
Ins 1,3,4-P_3 — Inositol 1,3,4-trisphosphate
Ins 1,4,5-P_3 — Inositol 1,4,5-trisphosphate
$InsP_4$ — Inositol tetrakisphosphate
PI — Phosphoinositide
PMA — Phorbol 12-myristate, 13-acetate

REFERENCES

1. **Berridge, M. J. and Irvine, R. F.,** Inositol trisphosphate, a novel second messenger in cellular signal transduction, *Nature,* 312, 315, 1984.
2. **Hokin, L. E.,** Receptors and phosphoinositide-generated second messengers, *Annu. Rev. Biochem.,* 55, 205, 1985.
3. **Bell, R. M.,** Protein kinase C activation by diacylglycerol second messengers, *Cell,* 45, 631, 1986.
4. **Nishizuka, Y.,** Studies and perspectives of protein kinase C, *Science,* 233, 305, 1986.
5. **Irvine, R. F., Dawson, R. M. C., and Freinkel, N.,** Stimulated phosphatidylinositol turnover: a brief appraisal, in *Contemporary Metabolism,* Vol. 2, Freinkel, N., Ed., Plenum Press, New York, 1982, 301.
6. **Irvine, R. F., Letcher, A. J., and Dawson, R. M. C.,** Phosphatidyl 4,5-bisphosphate phosphodiesterase and phosphomonoesterase activities of rat brain, *Biochem. J.,* 218, 177, 1984.
7. **Gilman, A. G.,** G proteins and dual control of adenylate cyclase, *Cell,* 36, 577, 1984.
8. **Birnbaumer, L., Codina, J., Mattera, R., Cerione, R. A., Hildebrandt, J. D., Sunyer, T., Rojas, F. J., Caron, M. G., Lefkowitz, R. J., and Iyengar, R.,** Regulation of hormone receptors and adenylate cyclase by guanine nucleotide binding N proteins, in *Recent Progress in Hormone Research,* Vol. 41, Greep, R. O., Ed., Academic Press, Orlando, FL, 1985, 41.
9. **Moss, J.,** Signal transduction by receptor-responsive guanyl nucleotide-binding proteins: modulation by bacterial toxin-catalyzed ADP-ribosylation, *Clin. Res.,* 35, 451, 1987.

10. **Harden, T. K., Tanner, L. I., Martin, M. W., Nakahata, N., Hughes, A. R., Evans, T., Masters, S. B., and Brown, J. H.,** Characteristics of two biochemical responses to stimulation of muscarinic cholinergic receptors, *Trends Pharmacol. Sci.,* 7 (Suppl.), 14, 1986.

11. **Hughes, A. R., Martin, M. W., and Harden, T. K.,** Pertussis toxin differentiates between two mechanisms of attenuation of cyclic AMP accumulation by muscarinic cholinergic receptors *Proc. Natl. Acad. Sci. U.S.A.,* 81, 5680, 1984.

12. **Tanner, L. I., Harden, T. K., Wells, J. N., and Martin, M. W.,** Purification of the phosphodiesterase regulated by muscarinic cholinergic receptors of 1321N1 human astrocytoma cells, *Mol. Pharmacol.,* 29, 455, 1986.

13. **Hokin, M. R. and Hokin, L. E.,** Effects of acetylcholine on phospholipids in pancreas, *J. Biol. Chem.,* 209, 549, 1954.

14. **Michell, R. H., Jafferji, S. S., and Jones, L. M.,** Receptor occupancy dose-response curve suggests that phosphatidylinositol breakdown may be intrinsic to the mechanism of the muscarinic cholinergic receptor, *FEBS Lett.,* 69, 1, 1976.

15. **Masters, S. B., Quinn, M. T., and Brown, J. H.,** Agonist-induced desensitization of muscarinic receptor-mediated calcium efflux without concomitant desensitization of phosphoinositide hydrolysis, *Mol. Pharmacol.,* 27, 325, 1985.

16. **Ambler, S. K., Solski, P. A., Brown, J. H., and Taylor, P.,** Receptor mediated inositol phosphate formation in relation to calcium mobilization: a comparison of two cell lines, *Mol. Pharmacol.,* 32, 376, 1987.

17. **Nakahata, N. and Harden, T. K.,** Regulation of inositol trisphosphate formation by muscarinic cholinergic and H1-histamine receptors on human astrocytoma cells: differential induction of desensitization by agonists, *Biochem. J.,* 241, 337, 1987.

18. **McDonough, P. M., Eubanks, J. H., and Brown, J. H.,** Desensitization and recovery of muscarinic and histaminergic calcium mobilization: evidence for a common hormone sensitive calcium store in 1321N1 astrocytoma cells, *Biochem. J.,* 249, 135, 1988.

19. **Glossman, H., Baukal, A., and Catt, K. J.,** Angiotensin II receptors in bovine adrenal cortex: modification of angiotensin II binding by guanyl nucleotides, *J. Biol. Chem.,* 249, 664, 1974.

20. **Rodbell, M.,** The role of hormone receptors and GTP-regulatory proteins in membrane transduction, *Nature,* 284, 17, 1980.

21. **Evans, T., Martin, M. W., Hughes, A. R., and Harden, T. K.,** Guanine nucleotide-sensitive high affinity binding of carbachol to muscarinic cholinergic receptors of 1321N1 astrocytoma cells is insensitive to pertussis toxin, *Mol. Pharmacol.,* 27, 32, 1985.

22. **Evans, T., Hepler, J. R., Masters, S. B., Brown, J. H., and Harden, T. K.,** Guanine nucleotide regulation of agonist binding to muscarinic cholinergic receptors: relation to efficacy of agonists for stimulation of phosphoinositide breakdown and Ca^{2+} mobilization, *Biochem. J.,* 232, 751, 1985.

23. **Haslam, R. J. and Davidson, M. M. L.,** Receptor-induced diacylglycerol formation in permeabilized platelets: possible role for a GTP-binding protein, *J. Receptor Res.,* 4, 605, 1984.

24. **Litosch, I., Wallis, C., and Fain, J. N.,** 5-Hydroxytryptamine stimulated inositol phosphate production in a cell-free system from blowfly salivary glands. Evidence for a role of GTP in coupling receptor activation to phosphoinositide breakdown, *J. Biol. Chem.,* 260, 5464, 1985.

25. **Cockcroft, S. and Gomperts, B. D.,** Role of a guanine nucleotide binding protein in the activtion of polyphosphoinositide phosphodiesterase, *Nature,* 314, 534, 1985.

26. **Orellana, S., Solski, P. A., and Brown, J. H.,** Guanosine 5'-O-(thiotriphosphate)-dependent inositol trisphosphate formation in membranes is inhibited by phorbol ester and protein kinase C, *J. Biol. Chem.,* 262, 1638, 1987.

27. **Hepler, J. R. and Harden, T. K.,** Guanine nucleotide-dependent pertussis toxin-insensitive stimulation of inositol phosphate formation by carbachol in a membrane preparation from human astrocytoma cells, *Biochem. J.,* 239, 141, 1986.

28. **Sternweis, P. C. and Gilman, A. G.,** Aluminum: a requirement for activation of the regulatory component of adenylate cyclase by fluoride, *Proc. Natl. Acad. Sci. U.S.A.,* 79, 4888, 1982.

29. **Higashijima, T., Ferguson, K. M., Sternweis, P. C., Ross, E. M., Smigel, M. D., and Gilman, A. G.,** The effect of activating ligands on the intrinsic fluorescence of guanine nucleotide-binding regulatory proteins, *J. Biol. Chem.,* 262, 752, 1987.

30. **Bigay, J., Deterre, P., Pfister, C., and Chabre, M.,** Fluoroaluminates activate transducin-GDP by mimicking the y-phosphate of GTP in its binding site, *FEBS Lett.,* 191, 181, 1985.

31. **Quilliam, L. A. and Brown, J. H.,** unpublished observation.

32. **Jones, L. G., Goldstein, D., and Brown, J. H.,** Guanine nucleotide-dependent inositol trisphosphate formation in chick heart cells, *Circ. Res.,* 62, 299, 1988.

33. **Merritt, J. E., Taylor, C. W., Rubin, R. P., and Putney, J. W.,** Evidence suggesting that a novel guanine nucleotide regulatory protein couples receptors to phospholipase C in exocrine pancreas, *Biochem. J.,* 236, 337, 1986.

34. **Dunlop, M. E. and Larkins, R. G.,** Muscarinic-agonist and guanine nucleotide activation of polyphosphoinositide phosphodiesterase in isolated islet-cell membranes, *Biochem. J.,* 240, 731, 1986.

35. **Nadler, E., Nijjar, M. S., and Oron, Y.,** Phosphoinositide breakdown in isolated rat parotid membranes, *FEBS Lett.,* 178, 278, 1984.

36. **Sasaguri, T., Hirata, M., Itoh, T., Koga, T., and Kuriyama, H.,** Guanine nucleotide binding protein involved in muscarinic responses in the pig coronary artery is insensitive to islet-activating protein, *Biochem. J.,* 239, 567, 1986.

37. **Jope, R. S., Casebolt, T. L., and Johnson, G. V. W.,** Modulation of carbachol-stimulated inositol phospholipid hydrolysis in rat cerebral cortex, *Neurochem. Res.,* 12, 693, 1987.

38. **Gonzales, R. A., and Crews, F. T.,** Guanine nucleotides stimulate production of inositol trisphosphate in rat cortical membranes, *Biochem. J.* 232, 799, 1985.

39. **Cockcroft, S.,** Polyphosphoinositide phosphodiesterase: regulation by a novel guanine nucleotide binding protein, G_p, *Trends Biochem. Sci.,* 12, 75, 1987.

40. **Ohta, H., Okajima, F., and Ui, M.,** Inhibition by islet-activating protein of a chemotactic peptide-induced early breakdown of inositol phospholipids and Ca^{2+} mobilization in guinea pig neutrophils, *J. Biol. Chem.,* 15780, 15771, 1985.

41. **Kikuchi, A., Kozawa, O., Kaibuchi, K., Katada, T., Ui, M., and Takai, Y.,** Direct evidence for involvement of a guanine nucleotide-binding protein in chemotactic peptide-stimulated formation of inositol biphosphate and trisphosphate in differentiated human leukemic (HL-60) cells, *J. Biol. Chem.,* 261, 11558, 1986.

42. **Nakamura, T. and Ui, M.,** Simultaneous inhibitions of inositol phospholipid breakdown, arachidonic acid release, and histamine secretion in mast cells by islet-activating protein, pertussis toxin: a possible involvement of the toxin-specific substrate in the Ca^{2+}-mobilizing receptor-mediated bio-signalling system, *J. Biol. Chem.,* 260, 3584, 1985.

43. **Verghese, M. W., Smith, C. D., and Snyderman, R.,** Potential role for a guanine nucleotide regulatory protein in chemo-attractant receptor mediated polyphosphoinositide metabolism, Ca^{2+} mobilization and cellular responses in leukocytes, *Biochem. Biophys. Res. Commun.,* 127, 450, 1985.

44. **Holian, A.,** Leukotriene B4 stimulation of phosphatidyl inositol turnover in macrophages and inhibition by pertussis toxin, *FEBS Lett.,* 201, 15, 1986.

45. **Pfeilschifter, J. and Bauer, C.,** Pertussis toxin abolishes angiotensin II-induced phosphoinositide hydrolysis and prostaglandin synthesis in rat mesangial cells, *Biochem. J.,* 236, 289, 1986.

46. **Johnson, R. M. and Garrison, J. C.,** Epidermal growth factor and angiotensin II stimulate formation of inositol-(1,4,5) and inositol-(1,3,4)-trisphosphate in hepatocytes: differential inhibition by pertussis toxin and phorbol-12-myristate-13-acetate, *J. Biol. Chem.,* 262, 17285, 1987.

47. **Xuan, Y-T., Su, T-F., Chang, K-J., and Watkins, W. D.,** A pertussis/cholera toxin sensitive G protein may mediate vasopressin-induced inositol phosphate formation in smooth muscle cell, *Biochem. Biophys. Res. Commun.,* 146, 898, 1987.

48. **Jackson, T. R., Hallam, T. J., Downes, C. P., and Hanley, M. R.,** Receptor coupled events in bradykinin action: Rapid production of inositol phosphates and regulation of cytosolic free Ca^{2+} in a neuronal cell line, *EMBO J.,* 6, 49, 1987.

49. **Paris, S. and Pouyssegur, J.,** Pertussis toxin inhibits thrombin-induced activation of phosphoinositide hydrolysis and Na^+/H^+ exchange in hamster fibroblasts, *EMBO J.,* 5, 55, 1986.

50. **Lo. W. W. Y. and Hughes, J.,** Pertussis toxin distinguishes between muscarinic receptor-mediated inhibition of adenylate cyclase and stimulation of phosphoinositide hydrolysis in Flow 9000 cells, *FEBS Lett.,* 220, 155, 1987.

51. **Martin, T. F. J., Bajjalieh, S. M., Lucas, D. O., and Kowalchyk, J. A.,** Thyrotropin-releasing hormone stimulation of polyphosphoinositide hydrolysis in GH_3 cell membranes is GTP dependent but insensitive to cholera or pertussis toxin, *J. Biol. Chem.,* 261, 10041, 1986.

52. **Masters, S. B., Martin, M. W., Harden, T. K., and Brown, J. H.,** Pertussis toxin does not inhibit muscarinic-receptor-mediated phosphoinositide hydrolysis or calcium mobilization, *Biochem. J.,* 227, 933, 1985.

53. **Jones, L. G. and Brown, J. H.,** Guanine nucleotide-regulated inositol polyphosphate production in adult rat cardiomyocytes, in *Biology of Isolated Adult Cardiac Myocytes,* Elsevier, New York, 257, 1988.

54. **Kelly, E., Rooney, T. A., and Nahorski, S. R.,** Pertussis toxin separates two muscarinic receptor-effector mechanisms in the striatum, *Eur. J. Pharmacol.,* 119, 129, 1985.

55. **Guillon, G., Balestre, M.-N., Mouillac, B., Berrada, R., and Kirk, C. J.,** Mechanisms of phospholipase C activation: a comparison with the adenylate cyclase system, *Biochimie,* 69, 351, 1987.

56. **Uhing, R. J., Prpic, V., Jiang, H., and Exton, J. H.,** Hormone-stimulated phosphoinositide breakdown in rat liver plasma membranes: role of guanine nucleotides and calcium, *J. Biol. Chem.,* 261, 2140, 1986.

57. **Jackson, T. R., Patterson, S. I., Wong, Y. H., and Hanley, M. R.,** Bradykinin stimulation of inositol phosphate and calcium responses is sensitive to pertussis toxin in NG115-401L neuronal cells, *Biochem. Biophys. Res. Commun.,* 148, 412, 1987.

58. **Murayama, T. and Ui, M.,** Receptor mediated inhibition of adenylate cyclase and stimulation of archidonic acid release in 3T3 fibroblasts: selective susceptibility to islet-activating protein, pertussis toxin, *J. Biol. Chem.,* 260, 7226, 1985.

59. **Schimmel, R. and Elliott, M.,** Pertussis toxin does not prevent α-adrenergic stimulated breakdown of phosphoinositides or respiration in brown adipocytes, *Biochem. Biophys. Res. Commun.,* 135, 823, 1986.

60. **Banga, H. S., Walker, R. K., Winberry, L. K., and Rittenhouse, S. E.,** Pertussis toxin can activate human platelets: comparative effects of holoprotein and its ADP-ribosylating S1 subunit, *J. Biol. Chem.,* 262, 14871, 1987.

61. **Pobiner, B. F., Hewlett, E. L., and Garrison, J. C.,** Role of N_i in coupling angiotensin receptors to inhibition of adenylate cyclase in hepatocytes, *J. Biol. Chem.,* 260, 16200, 1985.

62. **Huff, R. M. and Neer, E. J.,** Subunit interactions of native and ADP-ribosylated α39 and α41, two guanine nucleotide-binding proteins from bovine cerebral cortex, *J. Biol. Chem.,* 261, 1105, 1986.

63. **Orellana, S. A. and Brown, J. H.,** unpublished observations.

64. **Imboden, J. B., Shoback, D. M., Pattison, G., and Stobo, J. D.,** Cholera toxin inhibits the T-cell antigen receptor-mediated increases in inositol trisphosphate and cytoplasmic free calcium, *Proc. Natl. Acad. Sci. U.S.A.,* 83, 5673, 1986.

65. **Lo, W. W. Y. and Hughes, J.,** A novel cholera toxin sensitive G protein (Gc) regulating receptor-mediated phosphoinosiotide signalling in human pituitary clonal cells, *FEBS Lett.,* 220, 327, 1987.

66. **Biffen, M. and Martin, B. R.,** Polyphosphoinositide labeling in rat liver plasma membranes is reduced by preincubation with cholera toxin, *J. Biol. Chem.,* 262, 7744, 1987.

67. **Pike, L. J. and Eakes, A. T.,** Epidermal growth factor stimulates the production of phosphatidylinositol monophosphate and the breakdown of polyphosphoinositides in A431 cells, *J. Biol. Chem.,* 262, 1644, 1987.

68. **Orellana, S. A., Solski, P. A., and Brown, J. H.,** Phorbol ester inhibits phosphoinositide hydrolysis and calcium mobilization in cultured astrocytoma cells, *J. Biol. Chem.,* 260, 5236, 1985.

69. **Vicentini, L. M., Di Virgilio, F., Ambrosini, A., Pozzan, T., and Meldolesi, J.,** Tumor promoter phorbol 12-myristate 13-acetate inhibits phosphoinositide hydrolysis and cytosolic Ca^{2+} rise induced by the activation of muscarinic receptors in PC12 cells, *Biochem. Biophys. Res. Commun.,* 127, 310, 1985.

70. **Bradford, P. G. and Rubin, R. P.,** Guanine nucleotide regulation of phospholipase C activity in permeabilized rabbit neutrophils, *Biochem. J.,* 239, 97, 1986.

71. **Lucas, D. O., Bajjalieh, S. M., Kowalchyk, J. A., and Martin, T. F. J.,** Direct stimulation by thyrotropin-releasing hormone of polyphosphoinositide hydrolysis in GH_3 cell membranes by a guanine nucleotide-modulated mechanism, *Biochem. Biophys. Res. Commun.,* 132, 721, 1985.

72. **Smith, C. D., Uhing, R. J., and Snyderman, R.,** Nucleotide regulatory protein-mediated activation of phospholipase C in human polymorphonuclear leukocytes is disrupted by phorbol esters, *J. Biol. Chem.,* 262, 6121, 1987.

73. **Chuang, D.-M.,** Carbachol-induced accumulation inositol-1-phosphate in neurohybridoma NCB-20 cells: effects of lithium and phorbol esters, *Biochem. Biophys. Res. Commun.,* 136, 622, 1986.

74. **Taylor, C. W., Merritt, J. E., Rubin, R. P., and Putney, J. W.,** Effects of protein kinase C on receptor-regulated formation of isomers of inositol trisphosphate, *Biochem. Soc. Trans.,* 14, 1018, 1986.

75. **Bennet, C. F. and Crooke, S. T.,** Purification and characterization of a phosphoinositide-specific phospholipase C from guinea pig uterus: phosphorylation by protein kinase C in vivo, *J. Biol. Chem.,* 262, 13789, 1987.

76. **Molina y Vedia, L. M. and Lapetina, E. G.,** Phorbol 12, 13-dibutyrate and 1-oleyl-2-acetylglycerol stimulate inositol trisphosphate dephosphorylation in human platelets, *J. Biol. Chem.,* 261, 10493, 1986.

77. **Connolly, T. M., Lawing, W. J., Jr., and Majerus, P. W.,** Protein kinase C phosphorylates human platelet inositol trisphosphate 5′-phosphomonoesterase, increasing the phosphatase activity, *Cell,* 46, 951, 1986.

78. **Blackshear, P. J., Stumpo, D. J., Huang, J.-K., Nemenoff, R. A., and Spach, D. H.,** Protein kinase C-dependent and -independent pathways of proto-oncogene induction in human astrocytoma cells, *J. Biol. Chem.,* 262, 7774, 1987.

79. **Brown, K. D., Blakeley, D. M., Hamon, M. H., Lauri, M. S., and Corps, A. N.,** Protein kinase C-mediated negative-feedback inhibition of unstimulated and bombesin-stimulated polyphosphoinositide hydrolysis in Swiss-mouse 3T3 cells, *Biochem. J.,* 245, 631, 1987.

80. **Kariya, K.-I. and Takai, Y.,** Distinct functions of down-regulation-sensitive and -resistant types of protein kinase C in rabbit aortic smooth muscle cells, *FEBS Lett.,* 219, 119, 1987.

81. **Florio, V. A. and Sternweis, P. C.,** Reconstitution of resolved muscarinic cholinergic receptors with purified GTP-binding proteins, *J. Biol. Chem.,* 260, 3477, 1985.

82. **Haga, K., Haga, T., and Ichiyama, A.,** Reconstitution of the muscarinic acetylcholine receptor: guanine nucleotide sensitive high affinity binding of agonists to purified muscarinic receptors reconstituted with GTP-binding proteins (G_i and G_o), *J. Biol. Chem.,* 261, 10133, 1986.

83. **Haga, T., Berstein, G., Nishiyama, T., Uchiyama, H., and Haga, K.,** Biochemical studies of the muscarinic receptor, in *Neuroreceptors and Signal Transduction,* Plenum Press, New York, 1988.

84. **Banno, Y., Nagao, S., Katada, T., Nagata, K.-I., Ui, M., and Nozawa, Y.,** Stimulation by GTP-binding proteins (G$_i$, G$_o$) of partially purified phospholipase C activity from human platelet membranes, *Biochem. Biophys. Res. Commun.,* 146, 861, 1987.

85. **Ashkenazi, A., Winslow, J. W., Peralta, E. G., Peterson, G. L., Schimerlik, M. I., Capon, D. J., and Ramachandran, J.,** An M$_2$ muscarinic receptor subtype coupled to both adenylyl cyclase and phosphoinositide turnover, *Science,* 238, 672, 1987.

86. **Nomura, Y., Kaneko, S., Kato, K.-I., Yamagashi, S.-I., and Sugiyama, H.,** Inositol phosphate formation and chloride current responses induced by acetylcholine and serotonin through GTP-binding proteins in *Xenopus* oocyte after injection of rat brain messenger RNA, *Mol. Brain Res.,* 2, 113, 1987.

87. **Fukuda, K., Kubo, T., Akiba, I., Maeda, A., Mishina, M., and Numa, S.,** Molecular distinction between muscarinic acetylcholine receptor subtypes, *Nature,* 327, 623, 1987.

88. **Bonner, T. I., Buckley, N. J., Young, A. C., and Brann, M. R.,** Identification of a family of muscarinic acetylcholine receptor genes, *Science,* 237, 527, 1987.

89. **Peralta, E. G., Ashkenazi, A., Winslow, J. W., Smith, D. H., Ramachandran, J., and Capon, D. J.,** Specific expression of four human muscarinic acetylcholine receptors, *EMBO J.,* 6, 3923, 1987.

90. **Bourne, H. R. and Sullivan, K. A.,** Mammalian G proteins: models for *ras* proteins in transmembrane signaling?, *Cancer Surv.,* 5, 257, 1986.

91. **Preiss, J., Loomis, C. R., Bishop, W. R., Stein, R., Niedel, J. E., and Bell, R. M.,** Quantitative measurement of sn-1,2-diacylglycerols present in platelets, hepatocytes, and *ras*- and sistransformed normal rat kidney cells, *J. Biol. Chem.,* 261, 8597, 1986.

92. **Fleischman, L. F., Chahwala, S. B., and Cantley, L.,** *Ras*-transformed cells: altered levels of phosphatidylinositol 4,5-bisphosphate and catabolites, *Science,* 231, 407, 1986.

93. **Wakelam, M. J. O., Davies, S. A., Houslay, M. D., McKay, I., Marshall, C. J., and Hall, A.,** Normal p21 N^{-ras} couples bombesin and other growth factor receptors to inositol phosphate production, *Nature,* 323, 173, 1986.

94. **Chiarugi, V., Porciatti, F., Pasquali, F., and Bruni, P.,** Transformation of Balb/3T3 cells with EJ/T24/H^{-ras} oncogene inhibits adenylate cyclase response to β-adrenergic agonist while increases muscarinic receptor dependent hydrolysis of inositol lipids, *Biochem. Biophys. Res. Commun.,* 132, 900, 1985.

95. **Lacal, J. C., De La Pena, P., Moscat, J., Garcia-Barreno, P., Anderson, P. S., and Aaronson, S. A.,** Rapid stimulation of diacylglycerol production in *Xenopus* oocytes by microinjection of H-*ras* p21, *Science,* 238, 533, 1987.

96. **Heyworth, C. M., Whetton, A. D., Wong, S., Martin, B. R., and Houslay, M. D.,** Insulin inhibits the cholera toxin-catalysed ribosylation of a M$_r$ 25,000 protein in fat liver plasma membranes, *Biochem. J.,* 228, 593, 1985.

97. **Lapetina, E. G. and Reep, B. R.,** specific binding of [α-32P]GTP to cytosolic and membrane-bound proteins of human platelets correlates with the activation of phospholipase C, *Proc. Natl. Acad. Sci. U.S.A.,* 84, 2261, 1987.

98. **Bhullar, R. P. and Haslam, R. J.,** Detection of 23 — 27 kDa GTP-binding proteins in platelets and other cells, *Biochem. J.,* 245, 617, 1987.

99. **Schatzman, R. C., Grifo, J. A., Merrick, W. C., and Kuo, J. F.,** Phospholipid-sensitive, Ca^{2+} dependent protein kinase phosphorylates the β subunit of eukaryotic initiation factor 2 (eIF-2), *FEBS Lett.,* 159, 167, 1983.

100. **Katada, T., Gilman, A. G., Watanabe, Y., Bauer, S., and Jakobs, K. H.,** Protein kinase C phosphorylates the inhibitory guanine-nucleotide-binding regulatory component and apparently supresses its function in hormonal inhibition of adenylate cyclase, *Eur. J. Biochem.,* 151, 431, 1985.

101. **Orellana, S. A., Buss, J. E. and Brown, J. H.,** unpublished observation.

102. **Mumby, S.,** personal communication, 1987.

103. **Ohashi, Y. and Narumiya, S.,** ADP-ribosylation of M$_r$ 21,000 membrane protein by type D botulinum toxin, *J. Biol. Chem.,* 262, 1430, 1987.

104. **Matsuoka, I., Syuto, B., Kurihara, K., and Kubo, S.,** ADP-ribosylation of specific membrane proteins in pheochromocytoma and primary-cultured brain cells by botulinum neurotoxins type C and D, *FEBS Lett.,* 216, 295, 1987.

105. **Quilliam, L. A., Brown, J. H., and Buss, J. E.,** A 22 kDa *ras*-related G-protein is the substrate for an ADP-ribosyltransferase from *Clostridium botulinum, FEBS Lett.,* 238, 22, 1988.

106. **Fitzgerald, T. J., Uhing, R. J., and Exton, J. H.,** Solubilization of the vasopressin receptor from rat liver plasma membranes, *J. Biol. Chem.,* 261, 16871, 1986.

107. **Bielinski, D. F. and Fishman, J. B.,** Tissue-specific hormonal responsiveness may be regulated by receptor/G-protein interactions (abstr.), *J. Cell. Biol.,* 105(Suppl.), 16a, 1987.

108. **Ryu, S. H., Suh, P.-G., Lee, K.-Y., and Rhee, S. G.,** Bovine brain cytosol contains three immunologically distinct forms of inositolphospholipid-specific phospholipase C, *Proc. Natl. Acad. Sci. U.S.A.,* 84, 6649, 1987.
109. **Balcarek, J. M., Bennett, C. F., and Crooke, S. T.,** Molecular cloning of rat phosphoinositide-specific phospholipase C (abstr.), *J. Cell. Biol.,* 105(Suppl.), 16a, 1987.

*G Proteins in Phospholipase C-Independent
Calcium Movements*

Chapter 5

G-PROTEIN CONTROL OF VOLTAGE-DEPENDENT Ca²⁺ CURRENTS

Jürgen Hescheler and Walter Rosenthal

TABLE OF CONTENTS

I.   Introduction 80

II.   Measurement of Ca²⁺ Currents 80
       A.   Studied Cell Types and Applied Techniques 80
       B.   Evidence for at least Two Types of Ca²⁺ Currents 81

III.   Cyclic AMP-Independent Hormonal Modulations of Ca²⁺ Currents 82
       A.   Inhibition 82
       B.   Stimulation 82
       C.   Effects of cAMP on Ca²⁺ Currents 84

IV.   Evidence for the Involvement of G Proteins in Ca²⁺ Current
       Modulations 85
       A.   Effects of Guanine Nucleotide Analogs 85
       B.   Effects of Pertussis Toxin 86

V.   Possible Identity of G Proteins Involved in Ca²⁺ Current Modulations 87

VI.   Concluding Remarks 90

Acknowledgments 91

References 91

I. INTRODUCTION

In nearly all eukaryotic cells, there is a close correlation between changes in plasma membrane potential and cellular functions, the most important examples being electrose-cretory and electromechanical coupling. The link between electrical events and cellular responses is an alteration of Ca^{2+} permeability of the plasma membrane, which is low at negative membrane potentials, permitting a 10,000-fold concentration gradient sustained by energy-consuming mechanisms. Depolarization of the plasma membrane leads to an increase in Ca^{2+} permeability by activation of voltage-dependent Ca^{2+} channels. The resulting in crease in cytoplasmic Ca^{2+} serves as an intracellular signal and leads to an altered functional state of Ca^{2+}-dependent proteins.

Besides regulation by the membrane potential, Ca^{2+} channel activity is modulated by various extracellular signals, i.e., hormones and neurotransmitters. Consequently, extra-cellular signals may alter the response of cells to depolarization and, vice versa, change the membrane potential which depends on Ca^{2+} permeability (positive feedback).[1]

A well-established mechanism by which extracellular signals modulate Ca^{2+} channel activity is the phosphorylation of proteins closely related to or identical with the channel. The best-studied example is the cAMP-dependent phosphorylation of cardiac L-type Ca^{2+} channels, which represents the mechanism by which hormones and drugs (e.g., β-adrenergic agonists) exert positive inotropic effects.[2-5] Beta-adrenergic agonists raise cytosolic cAMP levels by stimulation of adenylate cylase via the cholera toxin-sensitive G protein, G_s.[6] On the other hand, inhibition of adenylate cyclase via a pertussis toxin-sensitive G protein of the G_i family by hormones like acetylcholine leads to decreased Ca^{2+} channel activity in cardiac myocytes, apparently by a reduction of cytosolic cAMP levels and subsequently reduced phosphorylation.[7-8] A phosphorylation of the protein forming the Ca^{2+} channel, proposed for cardiac cells, has so far only been demonstrated for dihydropyridine-binding sites purified from skeletal muscle. If reconstituted into phospholipid bilayers, these sites show Ca^{2+} channel activity which is increased by the catalytic subunit of the cAMP-dependent protein kinase and MgATP.[9] Ca^{2+} channels of hippocampal neurons may also be stimulated by cAMP- dependent phosphorylation.[10] In addition, cAMP-dependent phos-phorylation appears to be a prerequisite for Ca^{2+} channel activity in isolated membrane patches of GH_3 cells.[11] Other protein kinases, i.e., cGMP-dependent protein kinase in snail neurons[12] and protein kinase C in various cell types[13] have been assumed to influence Ca^{2+} channel activity. However, Ca^{2+} channel modulation by these protein kinases has not been shown to be due to direct phosphorylation of proteins forming Ca^{2+} channels.

Besides the principle described above, evidence is growing for the existence of another mechanism of Ca^{2+} channel modulations by extracellular signals. Typical for this mechanism is a close control of Ca^{2+} channels by receptor-activated G-proteins, without apparent involvement of protein kinases regulated by intracellular signal molecules such as cAMP, cGMP, or diacylglycerol.

In this chapter, we will describe examples of Ca^{2+} channel modulations that follow this latter principle. The G-protein control of Ca^{2+} channels raises two main questions: (1) which G proteins are involved in the action of Ca^{2+} channel-modulating receptor agonists and (2) what is the molecular mechanism by which G proteins control Ca^{2+} channel activity?

II. MEASUREMENT OF Ca^{2+} CURRENTS

A. STUDIED CELL TYPES AND APPLIED TECHNIQUES

Electrophysiological studies concerning the G-protein control of voltage-dependent Ca^{2+} channels have been performed with single cells obtained from tissues by collagenase-treat-ment or from established cell cultures, with isolated plasma membrane patches or with Ca^{2+}

channel-containing preparations reconstituted into artificial planar phospholipid bilayers. The patch-clamp technique, introduced by Neher and Sakmann,[14] provided a convenient method for measurement of whole-cell transmembranous currents as well as currents through single ion channels of cell-attached or isolated membrane patches or of ion channels incorporated into phospholipid bilayers. The whole-cell clamp configuration allows — besides superfusion of cells with various bathing solutions — the intracellular application of compounds via the patch pipette. Single-channel recording in the inside-out configuration provides free access to the cytoplasmic face of an isolated plasma membrane patch.

We studied Ca^{2+} currents and their hormonal modulation in three different cell lines, using the whole-cell clamp technique. Neuroblastoma $\times$ glioma hybrid (N$\times$G) cells (108CC15)[15] were used as a model for neuronal cells to examine the involvement of Ca^{2+} channels in the action of neurotransmitters known to inhibit transmitter release by activation of presynaptic receptors.[16] N$\times$G cells possess a large number of inhibitory opioid receptors of the δ-type which are activated by a synthetic peptide, D-Ala^2, D-Leu^5-enkephalin (DA-DLE). An adrenocortical cell line (Y1)[17] was used to test whether angiotensin II, which is a potent stimulator of steroid secretion, possibly exerts its effect by activation of Ca^{2+} channels. Finally, a pituitary cell line (GH$_3$)[18] appeared to be a suitable model since the cells possess receptors not only for secretion-stimulating but also for secretion-inhibiting hormones, e.g., gonadotropine-releasing hormone (LHRH) and somatostatin, respectively.

B. EVIDENCE FOR AT LEAST TWO TYPES OF Ca^{2+} CURRENTS

For determination of ion fluxes through voltage-dependent Ca^{2+} channels, Ba^{2+} is routinely used as charge carrier, and outward (K$^+$) currents are minimized by replacing K$^+$ for Cs$^+$ in the bathing (external) and pipette (internal) solutions. Fast Na$^+$ inward currents are blocked by tetrodotoxin or inactivated by appropriate voltage-clamp pulses. Under voltage-clamp conditions, cells are held at negative membrane potentials and Ca^{2+} currents are determined during depolarizing voltage-clamp pulses. In each of the tested cell types, depolarizing voltage-clamp pulses of 300 ms elicited Ba^{2+} inward (by definition negative) currents. At a holding potential of -80 mV, transient inward currents were evoked by test pulses to -30 mV. At a holding potential of -40 mV, a different type of current was evoked by test pulses to 0 mV; this current showed very little inactivation during the test pulse (Figure 1). Figure 2 shows superimposed current-voltage relations taken at various times after beginning of test pulses (see legend to Figure 2). When pulses were applied to a cell held at -100 mV, curves taken up to 30 ms after beginning of the pulse differed from those taken at the end of the pulse. At a less negative holding potential (-40 mV), curves were almost superimposable throughout the pulse and identical with those observed at the end of the pulse at a holding potential of -100 mV. Thus, all cell types exhibited fast and slowly inactivating inward currents in accordance with the identification of two types of Ca^{2+} currents in dorsal root ganglia of chick and rat,[19] N$\times$G cells,[20,21] isolated porcine[22] and bovine[23] glomerulosa cells, GH$_3$ cells,[24,25] and cardiac cells.[26]

Both types of inward currents were inhibited by Ni^{2+}, Cd^{2+}, and Co^{2+} but insensitive to tetrodotoxin. Organic Ca^{2+} channel agonists and antagonists mostly affected slowly inactivating currents of all cell types, whereas fast inactivating currents were not influenced by the organic compounds. According to the electrophysiological classification of Ca^{2+} channels into L-, T-, and N-type channels,[27] fast inactivating currents may mainly represent currents through T-type Ca^{2+} channels, whereas slowly inactivating currents may mainly represent currents through L- or N-type Ca^{2+} channels. According to the nomenclature of Carbone and Lux,[19] the identified fast and slowly inactivating inward currents may correspond to low and high threshold Ca^{2+} channels, respectively.

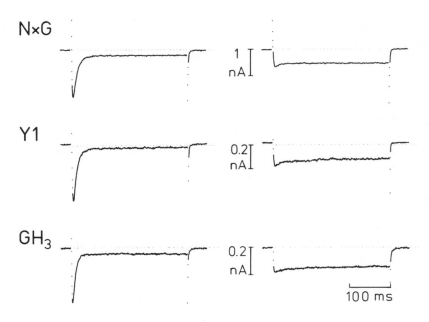

FIGURE 1. Ca^{2+} currents of $N \times G$, Y1, and GH_3 cells. Shown are current traces during test pulses (300 ms) from -80 to -30 (left panels) and -40 to 0 mV (right panels). For experimental details see References 22, 37, 38, and 40.

III. CYCLIC AMP-INDEPENDENT HORMONAL MODULATIONS OF Ca^{2+} CURRENTS

A. INHIBITION

Hormonal inhibition of Ca^{2+} currents has been observed in various cell types. Examples include dorsal root ganglia of chick,[28-30] rat,[31,32] and mouse,[33] sensory and sympathetic ganglia of chick,[34,35] rat sympathetic neurons,[36] $N \times G$ cells,[21,37,38] and a murine pituitary cell line, AtT20.[39] An example for inhibitory modulation of Ca^{2+} currents in an $N \times G$ cell is given in Figure 3. Superfusion of the cell with the δ-agonist, DADLE, caused an inhibition of Ca^{2+} currents by about 70% (see also Figure 5).[37,38] The effect was rapidly reversed by washing off the agonist. Similarly, the release-inhibiting hormone, somatostatin, caused a reversible inhibition of Ca^{2+} currents in a GH_3 cell (see Figure 3).[40]

There is evidence that, at least in neuronal cells, the Ca^{2+} current inhibited by receptor agonists represents ion fluxes through N-type Ca^{2+} channels.[21,33,36]

B. STIMULATION

There are few examples of hormones that do not activate adenylate cyclase (and, consequently, cAMP-dependent protein kinase) but appear to act by stimulation of Ca^{2+} channels. Biochemical evidence has been provided that angiotensin II-induced secretion of aldosterone involves activation of voltage-dependent Ca^{2+} channels. Both angiotensin II- and K^+-induced aldosterone secretions depend on extracellular Ca^{2+},[41] are inhibited by various Ca^{2+} channel blockers,[42-45] and are stimulated by the Ca^{2+} channel agonist, Bay K 8644.[44,46] In addition, angiotensin II and K^+ have been shown to stimulate[45] Ca^{2+} influx in a dihydropyridine-sensitive manner.[47,48] Figure 3 shows that angiotensin II stimulated slowly inactivating Ca^{2+} currents of Y1 cells (see also Figure 5).[22] Similarly to the action of inhibitory receptor agonists, the stimulation by angiotensin II occurred at concentrations ranging from 1 nM to 1 μM was observed within a few seconds and was rapidly reversed by removal of the hormone from the bath. Stimulation of Ca^{2+} currents by angiotensin II has also been observed in isolated porcine[22] and bovine adrenal glomerulosa cells.[23]

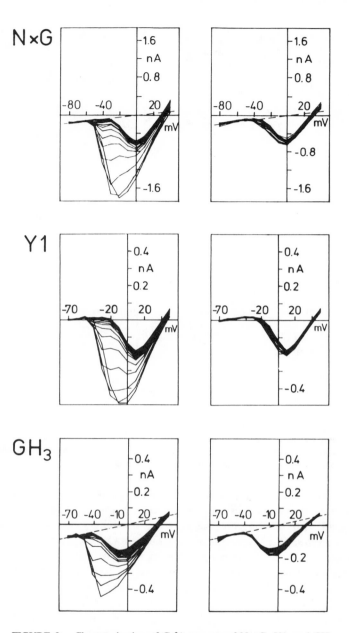

FIGURE 2. Characterization of Ca^{2+} currents of $N \times G$, Y1, and GH_3 cells. Inward currents were measured at holding potentials of -100 mV (left panels) or -40 mV (right panels). Shown are superimposed current-voltage relations. Currents were measured at 25 times within the 300 ms test pulse, i.e., 7, 9, 11, 15, 20, 25, 30, 35, 40, 45, 50, 60, 70, 80, 90, 100, 120, 140, 160, 180, 200, 220, 240, 260, 280, and 290 ms after beginning of the pulse. Leakage currents are indicated by interrupted lines. In the left panels, curves taken between 7 and 25 ms after beginning of the pulse showed a rapid decrease in the current amplitude at negative test potentials. For experimental details see References 22, 37, 38, and 40.

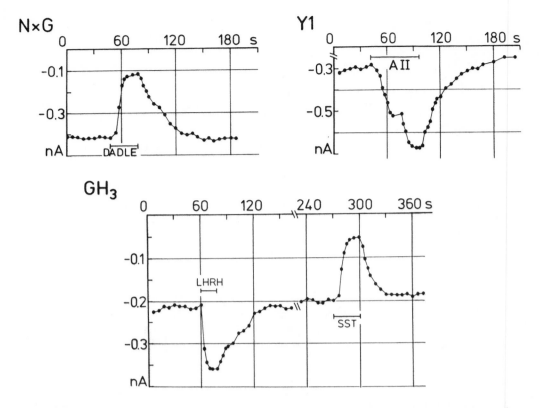

FIGURE 3.   Effect of various receptor agonists on Ca^{2+} currents in $N \times G$, Y1, and GH_3 cells. Shown are time courses of Ca^{2+} currents evoked by repetitive (0.2 Hz) voltage-clamp pulses from -40 to 0 mV. DADLE (1 μM), angiotensin II (0.1 μM, A II), LHRH (0.1 μM), and somatostatin (0.1 μM, SST) were present at the times indicated by the bars. For experimental details see References 22, 37, 38, and 40.

Another secretion-stimulating hormone, LHRH, has been suggested to act, at least in part, by activation of voltage-dependent Ca^{2+} channels. Biochemical support for this assumption is based on the finding that the response of pituitary cells to LHRH is reduced by removal of extracellular Ca^{2+} or by Ca^{2+} channel blockers of the dihydropyridine or the phenylalkylamine type[49,50] on one hand and potentiated by the Ca^{2+} channel agonist, Bay K 8644, on the other hand.[51] Figure 3 shows that LHRH stimulated slowly inactivating Ca^{2+} currents of a GH_3 cell.[40] As observed with the other receptor agonists, the effect was fast in onset and rapidly reversible. In the same cell, somatostatin caused a reversible inhibition of slowly inactivating Ca^{2+} currents (see above). When GH_3 cells were superfused with a mixture of LHRH and somatostatin at maximally effective concentrations of each hormone (1 μM), the inhibitory effect prevailed (not shown).

At present it is not clear which type of Ca^{2+} channel is stimulated by LHRH or angiotensin II. In both cell types tested, hormone-sensitive currents were partially sensitive towards organic channel agonists and antagonists of the dihydropyridine type, suggesting that fluxes through L-type Ca^{2+} channels contribute to these currents. Slowly inactivating Ca^{2+} currents of GH_3 cells are also inhibited by the snake venom, ω-conotoxin[52] which affects L- and N-type Ca^{2+} channels in neuronal tissues.[53] Thus, at present it is not clear whether hormones modulate N- or L-type Ca^{2+} channels in endocrine cells.

C. EFFECTS OF cAMP ON CA^{2+} CURRENTS

The stimulation of Ca^{2+} currents in cardiac and skeletal muscle by a cAMP-dependent phosphorylation[2-5,9] is the only well-established principle of Ca^{2+} channel modulation by

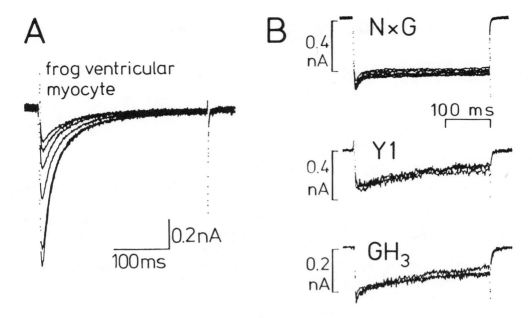

FIGURE 4. (Panel A) Stimulation of Ca^{2+} currents in frog ventricular myocytes by cAMP. Shown are super-imposed current traces recorded during voltage-clamp pulses from −40 to 0 mV. Traces were taken at the beginning and up to about 5 min after beginning of intracellular infusion. The pipette solution contained 100 μM cAMP. In this experiment, Ca^{2+} (1.8 mM) was used as a charge carrier.[7] (Panel B) Failure of cAMP to modulate Ca^{2+} currents in N × G, Y1, and GH_3 cells. Experimental conditions were as those described in panel A except that Ba^{2+} was used as charge carrier.[22,37,38,40]

extracellular signals. Therefore, it appears possible that the dual modulation of Ca^{2+} currents described above is a consequence of stimulation or inhibition of adenylate cyclase. This hypothesis is supported by the fact that extracellular signals known to inhibit voltage-dependent Ca^{2+} currents (e.g., adrenaline via α_2-receptors, GABA via $GABA_B$ receptors, dopamine via D_2 receptors, opioids via δ and κ-receptors, somatostatin) also inhibit adenylate cyclase. The concept is, however, not consistent with the observation that angiotensin II stimulates voltage-dependent Ca^{2+} currents, but inhibits adenylate cyclase in adrenocortical cells[54] and pituitary cells[55] including GH_3 cells.[40] In addition, LHRH stimulated voltage-dependent Ca^{2+} currents, but did not stimulate adenylate cyclase in membranes of GH_3 cells.[40] Therefore, a cAMP-dependent mechanism underlying hormonal stimulation of Ca^{2+} currents in endocrine cells appears to be unlikely.

Figure 4 shows that Ca^{2+} currents of frog myocytes were, as expected, stimulated by intracellularly applied cAMP, whereas Ca^{2+} currents of N × G, Y1, and GH_3 cells were not affected by the nucleotide. In addition, several groups have shown that inhibition of pituitary Ca^{2+} currents by somatostatin[39] and of neuronal Ca^{2+} currents by acetylcholine,[36] nor-adrenaline, and GABA[29] or by a GABA agonist[31] is not affected by intracellular infusion of cAMP.

IV. EVIDENCE FOR THE INVOLVEMENT OF G PROTEINS IN Ca^{2+} CURRENT MODULATIONS

A. EFFECTS OF GUANINE NUCLEOTIDE ANALOGS

The involvement of G protein in the Ca^{2+} current modulations described above were demonstrated by the use of GTP or GDP analogs that irreversibly activate or inactivate G proteins, respectively (see chapters on G proteins). Intracellular infusion of a GTP analog,

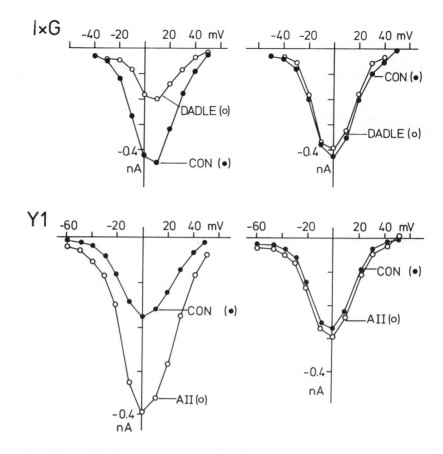

FIGURE 5.   Effects of pertussis toxin on modulations of Ca^{2+} currents in N × G (upper panels) and Y1 cells (lower panels). Holding potentials were −40 mV. Ca^{2+} currents were measured as maximal inward currents during test pulses in the absence and presence of the respective agonist applied at a concentration of 1 μM each. Left panels: current-voltage relations obtained from control cells. Right panels: current-voltage relations obtained from cells pretreated with pertussis toxin (100 ng/ml for 3 to 6 h). For experimental details see References 22, 37, and 38.

guanosine-5′-0-(3-thiotriphosphate) (GTPγS) at high concentrations (100 to 500 μM), drastically reduced Ca^{2+} currents in neuronal[29,31,36-38] and pituitary cells.[39] GTPγS at a low concentration (1 μM) had no effect on Ca^{2+} currents of N × G cells in the absence of DADLE but caused a marked *irreversible* inhibition of currents, if DADLE was present in the extracellular medium.[38] On the other hand, intracellular application of the GDP analog, guanosine-5′-0-(2-thiodiphosphate) (GDPβS), prevented the inhibition of Ca^{2+} currents by receptor agonists.[29,31,36-38] Thus, the effects of guanine nucleotide analogs on Ca^{2+} currents resemble their effects on hormone-sensitive adenylate cyclase.[56]

G proteins appear to be not only involved in the hormonal modulation of Ca^{2+} channels but also in the interaction of Ca^{2+} channels and Ca^{2+} channel ligands such as D-600, nifedipine, and diltiazem.[57] This assumption is based on the finding that in dorsal root ganglia the antagonistic effect of Ca^{2+} channel ligands is turned into an agonist effect by intracellular infusion of high concentrations (500 μM) of GTPγS.

B. EFFECTS OF PERTUSSIS TOXIN

The main exotoxin of *Bordetella pertussis* (pertussis toxin) prevents functional coupling of activated receptors and a number of G proteins (e.g., G_i-like G proteins and G_o) by ADP-ribosylation of G-protein α-subunits.[6,58] As is shown in Figure 5, hormonal inhibition of

Ca^{2+} currents in N × G cells was abolished by pretreatment of cells with pertussis toxin. In analogy to data obtained for inhibition of adenylate cyclase,[59] GTPγS-induced inhibition of Ca^{2+} currents in N × G cells was not affected by the toxin.[38] The suppression of the inhibitory effect of receptor agonists by pertussis toxin has also been described for neuronal[29] and pituitary cells.[39] In addition, pertussis toxin completely blocked the hormonal stimulation of Ca^{2+} currents in Y1 cells (see Figure 5). This observation is consistent with the biochemical finding that stimulation of Ca^{2+} influx by angiotensin II is abolished by the toxin.[48] Moreover, the bidirectional hormonal modulation of Ca^{2+} currents in GH_3 cells was completely suppressed in cells pretreated with the toxin (not shown).[40] The occurrence of both pertussis toxin-sensitive hormonal stimulation and inhibition of voltage-dependent Ca^{2+} currents in one cell type suggest that these opposite regulations are mediated by distinct G proteins.

It is unlikely that protein kinase C is involved in the Ca^{2+} current stimulation by angiotensin II and LHRH, since these hormones stimulate phosphatidylinositol 4,5-bisphosphate hydrolysis (causing activation of protein kinase C, see Chapter 3) via a pertussis toxin-insensitive G protein in adrenocortical[48,60] and pituitary cells.[61] In preliminary experiments, protein kinase C-activating phorbol esters did not affect Ca^{2+} currents of Y1 cells.

Participation of protein kinase C in inhibition of Ca^{2+} currents is also unlikely, since most of the Ca^{2+} current-inhibiting receptor agonists do not stimulate phosphatidylinositol 4,5-bisphosphate hydrolysis (e.g., somatostatin, opioids, $GABA_B$- and α_2-adrenoceptor-agonists). In addition, protein kinase C activators and inactivators did not influence Ca^{2+} currents of rat sympathetic neurons.[36]

V. POSSIBLE IDENTITY OF G PROTEINS INVOLVED IN Ca^{2+} CURRENT MODULATIONS

Neuronal cells including N × G cells (see Figure 7)[62] contain two types of pertussis toxin-sensitive G proteins: G_i-like G-proteins with molecular masses of their α-subunits ranging from 39.5 to 41 kDa[6,63] and G_o with a 39 kDa α-subunit.[64-66] The concentration of G_o in the central nervous system is several fold higher than that of G_i-like G proteins.[66,67]

We infused G proteins purified from porcine cerebral cortex into N × G cells which had become insensitive to DADLE following pretreatment with pertussis toxin. As shown in Figure 6A, intracellular infusion of a mixture of both G_i-like G proteins and G_o completely reconstituted the DADLE effect within 30 min. In order to elucidate which pertussis toxin-sensitive G protein is involved in functional coupling of opioid receptors and Ca^{2+} channels of N × G cells, individual G-proteins or isolated subunits were applied (Figure 6B). While at a concentration of about 4 nM in the patch pipette, both G_i-like G proteins and the α-subunit of G_o were able to restore the DADLE effect, only the α-subunit of G_o was effective at a ten times lower concentration.[37] High concentrations of the retinal (bovine) G-protein, transducin, and βγ-complex of G proteins purified from porcine cerebral cortex were not effective. The ability of G_o to mediate hormonal inhibition of neuronal Ca^{2+} currents has also been described by Miller and colleagues.[68] The high abundance of G_o in neuronal cells and its reconstituting effect at low concentrations suggest that it may play a role in functional coupling of inhibitory receptors and neuronal Ca^{2+} channels.

Data derived from reconstitution experiments concerning the identity of the pertussis toxin-sensitive G protein involved in stimulation of Ca^{2+} currents are not yet available. However, some conclusion can be drawn by relating hormonal effects on Ca^{2+} currents to the pattern of G-protein α-subunits in membranes of the cell types tested. For identification of G-protein α-subunits, we probed membranous fractions of N × G, Y1, and GH_3 cells with antisera raised against synthetic peptides corresponding to confined regions of G-protein α-subunits (Figure 7).[22,40] An antiserum raised against a sequence common to all G-protein

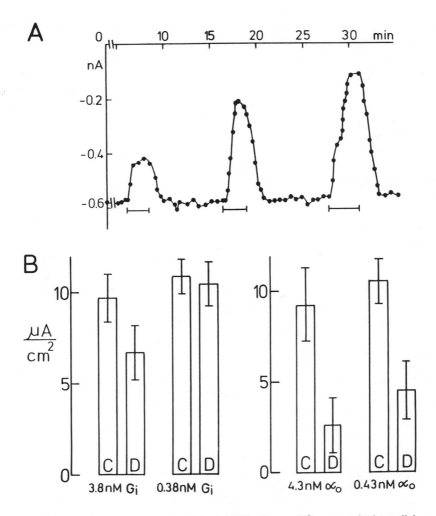

FIGURE 6.   (Panel A) Restoration of the DADLE effect on Ca^{2+} currents by intracellular infusion of G_i-like G proteins and G_o in a pertussis toxin-pretreated N × G cell. Currents were elicited by voltage-clamp pulses from −40 to 0 mV at a frequency of 0.2 Hz. DADLE (1 μM) was present at time intervals indicated by the bars. G proteins were purified from porcine brain.[67] The concentration of each G_i-like G proteins and G_o was 15 μM. (Adapted from Hescheler, J., Rosenthal, W., Trautwein, W., and Schultz, G., *Nature*, 325, 445, 1987.) (Panel B) Effect of intracellular infusion of various concentrations of G_i-like G proteins and of the α-subunit of G_o (α$_o$). Experimental procedures were as in panel A. Currents were taken about 15 min after disruption of membrane patches in the absence (C) and presence of 1 μM DADLE (D). Shown are mean values of peak current densities ± SD; the numbers of cells tested (n) varied between 4 to 5. (Adapted from Hescheler, J., Rosenthal, W., Wulfern, M., Tang, M., Yajima, M., Trautwein, W., and Schultz, G., *Adv. Second Messenger Phosphoprot. Res.*, 21, 165, 1988.)

α-subunits (α$_{common}$ peptide) reacted with membranous peptides of about 39 to 41 kDa M_r in all tested cell types and recognized the α-subunits of purified pertussis toxin-sensitive G proteins, i.e., G_i from human erythrocytes and porcine brain, G_o from procine brain, and the 39 kDa α-subunit of bovine transducin. Similarly an antiserum raised against a peptide common to α-subunits of G_i-like G proteins (α$_i$ peptide) reacted with peptides of about 40 kDa in membranes of all tested cell types; this serum recognized the α-subunits of G_i-like G proteins purified from human erythrocytes and porcine brain but not the α-subunits of

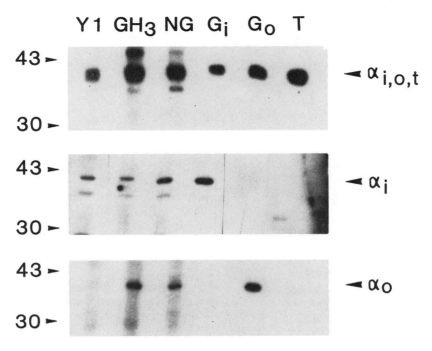

FIGURE 7.   Fig 5:G-protein α-subunits in membranous fractions of various cell types. Shown are autoradiographs of immunoblots obtained with antisera raised against synthetic peptides corresponding to confined regions of G-protein α-subunits. Membranous fractions of each indicated cell type (75 μg) or purified G proteins (1 μg with respect to G-protein α-subunit) were subjected to SDS-PAGE and transferred to nitrocellulose filters which were incubated with a 1:300 diluted antiserum raised against a peptide common to G-protein α-subunits (upper panel), a 1:300 diluted antiserum raised against a peptide specific for the α-subunit of G_i-like G proteins (middle panel), or a 1:300 diluted antiserum raised against a peptide specific for the α-subunit of G_o (lower panel). Immunoreactive peptides were identified by autoradiography following incubation of filters with ^{125}I-protein A. Figures on the left panel margins indicate relative molecular masses (in kDa). $\alpha_{i,o,t}$, α-subunits of G_i-like G proteins, G_o, and transducin (39 to 41 kDa). NG, N × G cells. (Adapted from Rosenthal, W., Hescheler, J., Hinsch, K.-D., Spicher, K., Trautwein, W., and Schultz, G., *EMBO J.*, 7, 1627, 1988.)

other purified G proteins. An antiserum raised against a peptide of the α-subunit of G_o (α_o peptide) specifically recognized peptides of about 39 kDa in membranes of GH_3 cells and N × G cells but not in Y1 cells; this serum reacted with the α-subunit of G_o but not with the α-subunits of other purified G proteins. The identification of the α-subunit of a G_i-like G protein in the absence of the α-subunit of G_o in membranes of Y1 cells indicate that the stimulatory effect of angiotensin II on Ca^{2+} currents is mediated by a G_i-like G protein and not by G_o. The results further indicate that GH_3 cells not only contain G_i-like G proteins but also the G protein, G_o, at concentrations similar to those found in cerebral cortex[66] and N × G cells (see Figure 7).[62] The high abundance of G_o in GH_3 cells — unique for peripheral cell type — may indicate that hormonal inhibition of nonneuronal Ca^{2+} channels is also mediated by G_o.

Recent findings by Yatani et al.[69] suggest that the cholera toxin-sensitive G protein, G_s, which mediates hormonal stimulation of adenylate cyclase, may stimulate cardiac Ca^{2+} channels without a cAMP-dependent intermediate step being involved. The physiological significance of this "direct" effect of G_s is not clear, since it has only been observed in cell-free systems (inside-out patches, artificial planar phospholipid bilayers) at high concentrations (20 to 100 pM) of the activated α-subunit of G_s.

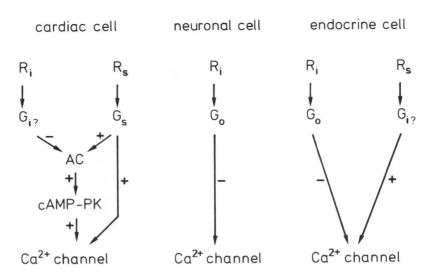

FIGURE 8.   Proposed mechanisms for hormonal modulations of voltage-dependent Ca^{2+} channels. R_i and R_s, inhibitory and stimulatory receptors, respectively; G_i?, not yet identified G_i-subtype; AC, adenylate cyclase; cAMP-PK, cAMP-dependent protein kinase. ($+$), stimulation; ($-$), inhibition. For further explanations see text.

VI. CONCLUDING REMARKS

Hormonal modulations of voltage-dependent Ca^{2+} currents in vertebrate cells appear to follow at least two G-protein-involving mechanisms (Figure 8). If protein kinases stimulated by intracellular signals are involved (e.g., cAMP-dependent protein kinase in the case of cardiac Ca^{2+} channels), G proteins participate indirectly in channel modulation as they mediate hormonal regulation of intracellular signal-generating enzymes. On the other hand, there is evidence for a close stimulatory and inhibitory control of Ca^{2+} channels by G proteins which appear to be membrane confined as there is no evidence for the involvement of cytosolic signal molecules or of protein kinases activated by these signal molecules.

The abundance of G_o in neuronal cells and in pituitary cells as well as experimental evidence derived from reconstitution experiments suggests that this G protein is involved in hormonal inhibition of Ca^{2+} channels in these two cell types. The pertussis-toxin-sensitive hormonal stimulation of Ca^{2+} currents in a cell line (Y1) that possesses a G_i-like G protein but lacks G_o argues for a role of a G_i-like G protein in hormonal stimulation of Ca^{2+} channels in this and possibly other endocrine cells. A second stimulatory, membrane-confined G-protein control of cardiac Ca^{2+} channels has been proposed by Yatani et al.,[69] describing a stimulatory effect of the isolated α-subunit of G_s in isolated membrane patches from cardiac ventricular cells.

One reason for these multiple principles of Ca^{2+} channel regulation may be the existence of several forms of Ca^{2+} channel proteins differing in their primary structure, subunit composition or posttranslational modifications including, e.g., phosphorylation, glycosylation, or acylation. Another reason may be a tissue-specific "environment" of ubiquitous Ca^{2+} channel proteins. The latter assumption is supported by the finding that ω-conotoxin, a snake venom, blocks L- and N-type Ca^{2+} currents in neuronal cells but not in cardiac myocytes.[53]

In contrast to the G-protein-mediated control of enzymatic effectors such as adenylate cyclase and retinal cGMP phosphodiesterase,[6] a direct, reversible interaction of G-proteins with Ca^{2+} channels has not yet been shown. Although the involvement of cytosolic signal molecules is unlikely, so far unknown membranous components may be required for functional coupling of G proteins and Ca^{2+} channels. It remains possible that G proteins directly regulate membrane-associated, not yet identified protein kinases or phosphatases, which control channel activity by phosphorylation or dephosphorylation of the channel protein. Alternatively, G proteins themselves may modify the channel protein and, as a consequence, its function by an inherent protein kinase or phosphatase activity.

The apparently close control of Ca^{2+} channels by G-proteins may be similar to that of K^+ channels in cardiac pacemaker cells and GH_3 cells. Stimulation of K^+ currents by acetylcholine[70,71] or somatostatin[72] in the respective cell type is independent of cyclic nucleotides but abolished by pretreatment of cells with pertussis toxin. In both systems, the activated α-subunit of G_i-like G protein is sufficient to enhance channel activity if applied to the cytosolic face of isolated membrane patches.[72-75] Therefore, the active G_i-subtype purified from erythrocytes or liver, possibly identical with G_{i3},[6,63] has been termed "G_K". It is an intriguing task to find out whether "G_K" is identical with the G_i-like G protein that confers pertussis toxin-sensitive hormonal stimulation to voltage-dependent Ca^{2+} channels or whether another member of the G_i family serves this function.

ACKNOWLEDGMENTS

We are grateful to Rosemarie Krüger for help in typing the manuscript. Work reported herein was supported by the Deutsche Forschungsgemeinschaft, SFB 246, and the Fonds der Chemischen Industrie.

REFERENCES

1. **Hille, B.**, *Ionic Channels of Excitable Membranes*, Sinauer Associates, Sunderland, 1984, 90.
2. **Reuter, H.**, Calcium channel modulation by neurotransmitters, enzymes and drugs, *Nature*, 301, 569, 1983.
3. **Kameyama, M., Hofmann, F., and Trautwein, W.**, On the mechanism of β-adrenergic regulation of the Ca channels in the guinea-pig heart, *Pflügers Arch.*, 405, 285, 1985.
4. **Kameyama, M., Hescheler, J., Hofmann, F., and Trautwein, W.**, Modulation of Ca current during the phosphorylation cycle in the guinea pig heart, *Pflügers Arch.*, 407, 123, 1986.
5. **Hofmann, F., Nastainczyk, W., Röhrkasten, A., Schneider, T., and Sieber, M.**, Regulation of the L-type calcium channel, *TIPS*, 8, 393, 1987.
6. **Graziano, M. P. and Gilman, A. G.**, Guanine nucleotide-binding regulatory proteins: mediators of transmembrane signaling, *TIPS*, 8, 478, 1987.
7. **Hescheler, J., Kameyama, M., and Trautwein, W.**, On the mechanism of muscarinic inhibition of the cardiac Ca current, *Pflügers Arch.*, 407, 182, 1986.
8. **Fischmeister, R. and Hartzell, C.**, Mechanism of acetylcholine on calcium current in single cells from frog ventricle, *J. Physiol. (London)*, 376, 183, 1986.
9. **Flockerzi, V., Oeken, H.-J., Hofmann, F., Pelzer, D., Cavalié, A., and Trautwein, W.**, Purified dihydropyridine-binding site from skeletal muscle t-tubules is a functional calcium channel, *Nature*, 323, 66, 1986.
10. **Gray, R. and Johnston, D.**, Noradrenaline and β-adrenoceptor agonists increase activity of voltage-dependent calcium channels in hippocampal neurons, *Nature*, 327, 620, 1987.
11. **Armstrong, D. and Eckert, R.**, Voltage-activated calcium channels that must be phosphorylated to respond to membrane depolarization, *Proc. Natl. Acad. Sci. U.S.A.*, 84, 2518, 1987.
12. **Paupardin-Tritsch, D., Hammond, C., Gerschenfeld, H. M., Nairn, A. C., and Greengard, P.**, cGMP-dependent protein kinase enhances Ca^{2+} current and potentiates the serotonin-induced Ca^{2+} current increase in snail neurones, *Nature*, 232, 536, 1986.

13. **Kaczmarek, L. K.,** The role of protein kinase C in the regulation of ion channels and neurotransmitter release, *TINS,* 10, 30, 1987.

14. **Hamill, O. P., Marty, A., Neher, E., Sakman, B., and Sigworth, F. J.,** Improved patch-clamp techniques for high-resolution current recording from cells and cell-free membrane patches, in *Single-Channel Recording,* Sakman, B. and Neher, E., Eds., Plenum Press, New York, 1981, 481.

15. **Hamprecht, B., Glaser, T., Reiser, G., Bayer, E., and Propst, F.,** Culture and characteristics of hormone-responsive neuroblastoma × glioma hybrid cells, *Methods Enzymol.,* 109, 316, 1985.

16. **Starke, K.,** Presynaptic α-autoreceptors, *Rev. Physiol. Biochem. Pharmacol.,* 107, 73, 1987.

17. **Schimmer, B. P.,** Isolation of ACTH-resistant Y1 adrenal tumor cells, *Methods Enzymol.,* 109, 350, 1985.

18. **Tashjian, A. H., Jr.,** Clonal strains of hormone-producing pituitary cells, *Methods Enzymol.,* 58, 527, 1979.

19. **Carbone, E. and Lux, H. D.,** A low voltage-activated, fully inactivating Ca channel in vertebrate sensory neurones, *Nature,* 310, 501, 1984.

20. **Hering, S., Bodewei, R., Schubert, B., Rhode, K., and Wollenberger, A.,** A kinetic analysis of the inward calcium current in 108CC15 neuroblastoma × glioma hybrid cells, *Gen. Physiol. Biophys.,* 4, 129, 1985.

21. **Tsunoo, A., Yoshii, M., and Narahashi, T.,** Block of calcium channels by enkephalin and somatostatin in neuroblastoma-glioma hybrid NG108-15 cells, *Proc. Natl. Acad. Sci. U.S.A.,* 83, 9832, 1986.

22. **Hescheler, J., Rosenthal, W., Hinsch, K.-D., Wulfern, M., Trautwein, W., and Schultz, G.,** Angiotensin II-induced stimulation of voltage-dependent Ca^{2+} currents in an adrenal cortical cell lines, *EMBO J.,* 7, 619, 1988.

23. **Cohen, C. J., McCarthy, R. T., Barrett, P. Q., and Rasmussen, H.,** Two populations of Ca channels in bovine glomerulosa cells, *Biophysical J.,* 53, 557a, 1987.

24. **Armstrong, C. M. and Mattison, D. R.,** Two distinct populations of calcium channels in a clonal line of pituitary cells, *Science,* 227, 65, 1985.

25. **DeRiemer, S. A. and Sakman, B.,** Two calcium currents in normal rat anterior pituitary cells identified by a plaque technique, *Exp. Brain Res.,* 14, 139, 1985.

26. **Nilius, B., Hess, P., Lansman, J. B., and Tsien, R. W.,** A novel type of cardiac calcium channel in ventricular cells, *Nature,* 316, 443, 1985.

27. **Nowycky, M. C., Fox, A. P., and Tsien, R. W.,** Three types of neuronal calcium channel with different calcium agonist sensitivity, *Nature,* 316, 440, 1985.

28. **Dunlap, K. and Fischbach, G. D.,** neurotransmitters decrease the calcium conductance activated by depolarization of embryonic chick sensory neurones. *J. Physiol.,* 317, 519, 1981.

29. **Holz, G. G., IV, Rane, S. G., and Dunlap, K.,** GTP-binding proteins mediated transmitter inhibition of voltage-dependent calcium channels, *Nature,* 319, 670, 1986.

30. **Forscher, P., Oxford, G. S., and Schulz, D.,** Noradrenaline modulates calcium channels in avian dorsal root ganglion cells through tight receptor-channel coupling, *J. Physiol.,* 379, 131, 1986.

31. **Scott, R. H. and Dolphin, A. C.,** Regulation of calcium currents by a GTP analogue: potentiation of (−)-baclofen-mediated inhibition, *Neurosci. Lett.,* 69, 59, 1986.

32. **Dolphin, A. C., Forda, S. R., and Scott, R. H.,** Calcium-dependent currents in cultured rat dorsal root ganglion neurones are inhibited by an adenosine analogue, *J. Physiol.,* 373, 47, 1986.

33. **Gross, R. A. and Macdonald, R. L.,** Dynorphin A selectively reduces a large transient (N-type) calcium current of mouse dorsal root ganglion neurons in cell culture, *Proc. Natl. Acad. Sci. U.S.A.,* 84, 5469, 1987.

34. **Deisz, R. A. and Lux, H. D.,** γ-Aminobutyric acid-induced depression of calcium currents of chick sensory neurons, *Neurosci. Lett.,* 56, 205, 1985.

35. **Marchetti, C., Carbone, E., and Lux, H. D.,** Effects of dopamine and noradrenaline on Ca channels of cultured sensory and sympathetic neurons of chick, *Pflügers Arch.,* 406, 104, 1986.

36. **Wanke, E., Ferroni, A., Malgaroli, A., Ambrosini, A., Pozzan, T., and Meldolesi, J.,** Activation of a muscarinic receptor selectively inhibits a rapidly inactivated Ca^{2+} current in rat sympathetic neurons, *Proc. Natl. Acad. Sci. U.S.A.,* 84, 4313, 1987.

37. **Hescheler, J., Rosenthal, W., Trautwein, W., and Schultz, G.,** The GTP-binding protein, G_o, regulates neuronal calcium channels, *Nature,* 325, 445, 1987.

38. **Hescheler, J., Rosenthal, W., Wulfern, M., Tang, M., Yajima, M., Trautwein, W., and Schultz, G.,** Involvement of the guanine nucleotide-binding protein, N_o, in the inhibitory regulation of neuronal calcium channels, *Adv. Second Messenger Phosphoprot. Res.,* 21, 165, 1988.

39. **Lewis, D. L., Weight, F. F., and Luini, A.,** A guanine nucleotide-binding protein mediates the inhibition of voltage-dependent calcium current by somatostatin in a pituitary cell line, *Proc. Natl. Acad. Sci. U.S.A.,* 83, 9035, 1986.

40. **Rosenthal, W., Hescheler, J., Hinsch, K.-D., Spicher, K., Trautwein, W., and Schultz, G.,** Cyclic AMP-independent, dual regulation fo voltage-dependent Ca^{2+} currents by LHRH and comatostatin in a pituitary cell line, *EMBO J.,* 7, 1627, 1988.

41. **Fakunding, J. L., Chow, R., and Catt, K. J.,** The role of calcium in the stimulation of aldosterone production by adrenocorticotropin, angiotensin II, and potassium in isolated glomerulosa cells, *Endocrinology,* 105, 327, 1979.

42. **Fakunding, J. L. and Catt, K. J.,** Dependence of aldosterone stimulation in adrenal glomerulosa cells on calcium uptake: effects of lanthanum and verapamil, *Endocrinology,* 107, 1345, 1980.

43. **Foster, R., Lobo, M. V., Rasmussen, H., and Marusic, E. T.,** Calcium: its role in the mechanism of action of angiotensin II and potassium in aldosterone production, *Endocrinology,* 109, 2196, 1981.

44. **Kojima, K., Kojima, I., and Rasmussen, H.,** Dihydropyridine calcium agonist and antagonist effects on aldosterone secretion, *Am. J. Physiol.,* 247, 1984.

45. **Aguilera, G. and Catt, K. J.,** Participation of voltage-dependent calcium channels in the regulation of adrenal glomerulosa function by angiotensin II and potassium, *Endocrinology,* 118, 112, 1986.

46. **Hausdorff, W. P., Aguilera, G., and Catt, K. J.,** Selective enhancement of angiotensin II- and potassium-stimulated aldosterone secretion by the calcium channel agonist BAY K 8644, *Endocrinology,* 118, 869, 1986.

47. **Kojima, I., Kojima, K., and Rasmusen, H.,** Characteristics of agiotensin II-, K$^+$ and ACTH-induced calcium influx in adrenal glomerulosa cells. Evidence that angiotensin II, K$^+$ and ACTH may open a common calcium channel, *J. Biol. Chem.,* 260, 9171, 1985.

48. **Kojima, I., Shibata, H., and Ogata, E.,** Pertussis toxin blocks angiotensin II-induced calcium influx but not inositol trisphosphate production in adrenal glomerulosa cell, *FEBS Lett.,* 204, 347, 1986.

49. **Marian, J. and Conn, P. M.,** Gonadotropin releasing hormone stimulation of cultured pituitary cells requires calcium, *Mol. Pharmacol.,* 16, 196, 1979.

50. **Conn, P. M., Rogers, D. C., and Seay, S. G.,** Structure-function relationship of calcium ion channel antagonists at the pituitary gonadotroph, *Endocrinology,* 113, 1592, 1983.

51. **Chang, J. P., McCoy, E. E., Graeter, J., Tasaka, K., and Catt, K. J.,** Participation of voltage-dependent calcium channels in the action of gonadotropin-releasing hormone, *J. Biol. Chem.,* 261, 9105, 1986.

52. **Suzuki, N. and Yoshioka, T.,** Differential blocking action of synthetic ω-conotoxin on components of Ca^{2+} channel current in clonal GH3 cells, *Neurosci. Lett.,* 75, 235, 1987.

53. **McCleskey, E. W., Fox, A. P., Feldman, D. H., Cruz, L. J., Olivera, B. M., Tsien, R. W., and Yoshikami, D.,** ω-Conotoxin: direct and persistent blockade of specific types of calcium channels in neurons but not muscle, *Proc. Natl. Acad. Sci. U.S.A.,* 84, 4327, 1987.

54. **Woodcock, E. A. and Johnson, C. I.,** Inhibition of adenylate cyclase in rat adrenal glomerulosa cells by angiotensin II, *Endocrinology,* 115, 337, 1984.

55. **Enjalbert, A., Sladeczek, F., Guillon, G., Bertrand, P., Shu, C., Epelbaum, J., Garcia-Sainz, A., Jard, S., Lombard, C., Kordon, C., and Bockaert, J.,** Angiotensin II and dopamine modulate both cAMP and inositol phosphate productions in anterior pituitary cells. Involvement in prolactin secretion, *J. Biol. Chem.,* 261, 4071, 1986.

56. **Birnbaumer, L., Codina, J., Mattera, R., Cerione, R. A., Hildebrandt, J. D., Sunyer, T., Rojas, F. J., Caron, M. G., Lefkowitz, R. J., and Iyengar, R.,** Regulation of hormone receptors and adenylyl cyclases by guanine nucleotide binding N proteins, *Rec. Prog. Hormone Res.,* 41, 41, 1985.

57. **Scott, R. H. and Dolphin, A. C.,** Activation of a G protein promotes agonist responses to calcium channel ligands, *Nature,* 330, 760, 1988.

58. **Ui, M., Katada, T., Murayama, T., Kurose, H., Yajima, M., Tamura, M., Nakamura, T., and Nogimori, K.,** Islet-activating protein, pertussis toxin: a specific uncoupler of receptor-mediated inhibition of adenylate cyclase, *Adv. Cyclic Nucleotide Protein Phosphorylation Res.,* 17, 145, 1984.

59. **Jakobs, K. H., Aktories, K., and Schultz, G.,** Mechanisms and components involved in adenylate cyclase inhibition by hormones, *Adv. Cyclic Nucleotide Protein Phosphorylation Res.,* 17, 135, 1984.

60. **Enyedi, P., Mucsi, I., Hunyady, L., Catt, K. J., and Spät, A.,** The role of guanyl nucleotide-binding proteins in the formation of inositol phosphates in adrenal glomerulosa cells, *Biochem. Biophys. Res. Commun.,* 140, 941, 1986.

61. **Naor, Z., Azrad, A., Limor, R., Zakut, H., and Lotan, M.,** Gonadotropin-releasing hormone activates a rapid Ca^{2+}-independent phosphodiester hydrolysis of polyphosphoinositides in pituitary gonadotrophs, *J. Biol. Chem.,* 261, 12506, 1986.

62. **Milligan, G., Gierschik, P., Spiegel, A. M., and Klee, W. A.,** The GTP-binding regulatory proteins of neuroblastoma × glioma, NG108-15 and glioma, C6, cells, *FEBS Lett.,* 195, 225, 1986.

63. **Suki, W. N., Abramowitz, J., Mattera, R., Codina, J., and Birnbaumer, L.,** The human genome encodes at least three non-allelic G proteins with α$_i$-type subunits, *FEBS Lett.,* 220, 187, 1987.

64. **Sternweis, P. C. and Robishaw, J. D.,** Isolation of two proteins with high affinity for guanine nucleotides from membranes of bovine brain, *J. Biol. Chem.,* 259, 13806, 1984.

65. **Neer, E. J., Lok, J. M., and Wolf, L. G.,** Purification and properties of the inhibitory guanine nucleotide regulatory unit of brain adenylate cyclase, *J. Biol. Chem.,* 259, 14222, 1984.

66. **Huff, R. M., Axton, J. M., and Neer, E. J.,** Physical and immunological characterization of a guanine nucleotide-binding protein purified from bovine cerebral cortex, *J. Biol. Chem.,* 260, 10864, 1985.

67. **Rosenthal, W., Koesling, D., Rudloph, U., Kluess, C., Pallast, M., and Schultz, G.,** Identification and characterization of the 35-kDa β subunit of guanine nucleotide-binding proteins by an antiserum raised against transducin, *Eur. J. Biochem.,* 158, 255, 1986.

68. **Miller, R. J.,** G proteins flex their muscles, *TINS,* 11, 3, 1988.

69. **Yatani, A., Codina, J., Imoto, Y., Reeves, J. P., Birnbaumer, L., and Brown, A. M.,** A G protein directly regulates mammalian cardiac calcium channels, *Science,* 238, 1288, 1987.

70. **Pfaffinger, P. J., Martin, J. M., Hunter, D. D., Nathanson, N. M., and Hille, B.,** GTP-binding proteins couple cardiac muscarinic receptors to a K channel, *Nature,* 317, 536, 1985.

71. **Breitwieser, G. and Szabo, G.,** Uncoupling of cardiac muscarinic and β-adrenergic receptors from ion channels by a guanine nucleotide analogue, *Nature,* 317, 538, 1985.

72. **Yatani, A., Codina, J., Sekura, R. D., Birnbaumer, L., and Brown, A. M.,** Reconstitution of somatostatin and muscarinic receptor mediated stimulation of K^+ channels by isolated G_k protein in clonal rat anterior pituitary cell membranes, *Mol. Endocrinol.,* 1, 283, 1987.

73. **Yatani, A., Codina, J., Brown, A. M., and Birnbaumer, L.,** Direct activation of mammalian atrial muscarinic potassium channels by GTP regulatory protein G_k, *Science,* 235, 207, 1987.

74. **Codina, J., Grenet, D., Yatani, A., Birnbaumer, L., and Brown, A. M.,** Hormonal regulation of pituitary GH_3 cell K^+ channels by G_k is mediated by its α-subunit, *FEBS Lett.,* 216, 104, 1987.

75. **Codina, J., Yatani, A., Grenet, D., Brown, A. M., and Birnbaumer, L.,** The α-subunit of the GTP binding protein G_k opens atrial potassium channels, *Science,* 236, 442, 1987.

Chapter 6

INOSITOL PHOSPHATE- AND GUANINE NUCLEOTIDE-ACTIVATED CALCIUM TRANSLOCATION WITHIN CELLS

Donald L. Gill, Julienne M. Mullaney, Tarun K. Ghosh, and Fahmy I. Tarazi

TABLE OF CONTENTS

I. Intracellular Calcium Signaling 96
    A. Calcium-Regulatory Organelles... 96
    B. Inositol Phosphate-Activated Calcium Movements 97

II. Guanine Nucleotide-Activated Calcium Release 98
    A. Original Observations 98
    B. Guanine Nucleotide Sensitivity and Specificity......................... 98
    C. Cellular and Subcellular Specificity 100
    D. Mechanism of Calcium Translocation:.................................. 101
        1. Reversibility of Action of GTP 101
        2. Membrane Interactions... 101

III. Relationship between IP$_3$- and GTP-Activated Calcium Release 103
    A. Distinctions in Mechanism of Action of IP$_3$ and GTP 103
    B. IP$_3$- and GTP-Releasable Calcium Pools 105

IV. Mechanism of Action of GTP on Calcium Movements 106
    A. GTP-Activated Calcium Uptake 107
        1. Oxalate Effects 107
        2. Identity of GTP-Activated Calcium Uptake and Release 108
    B. Clues to the Action of GTP in Mediating Calcium Movements 109
    C. The GTP-Activated Transmembrane Calcium ''Conveyance''
        Model 111
    D. GTP-Induced Loading of the IP$_3$-Releasable Pool 113

V. Conclusions and Scheme for the Actions of IP$_3$ and GTP..................... 116

References.. 117

I. INTRACELLULAR CALCIUM SIGNALING

The central roles of Ca^{2+} in the regulation of cellular function and as an intracellular mediator of receptor-activated signaling are now well recognized. The recent elucidation of the mechanisms coupling cell-surface receptors to Ca^{2+} mobilization in cells based on the early observations of the Hokins[1] has established in principle the relationship between receptor-induced phosphoinositide breakdown and inositol phosphate-mediated Ca^{2+} release.[2-5] However, it should be noted that Ca^{2+} regulation within cells is a complex event involving an array of distinct transport mechanisms, located in a number of discrete organelles, and influenced by numerous intracellular regulatory systems. It is the purpose of this chapter to review some intriguing recent developments concerning the control of intracellular Ca^{2+} movements and their possible relationship to what has been ascertained on the processes that mediate Ca^{2+} signaling events within cells. In this section, certain of the characteristics of Ca^{2+} regulatory organelles and their role in Ca^{2+} signaling are considered.

A. CALCIUM-REGULATORY ORGANELLES

The movements of Ca^{2+} ions across membranes within cells are controlled by a number of distinct types of active or passive transport mechanisms.[6] The cytosol of most mammalian cells contains approximately 0.1 μM free Ca^{2+} under resting conditions as compared to the low millimolar free Ca^{2+} concentration outside cells. This 10,000-fold gradient of free Ca^{2+} across the plasma membrane is actively maintained via ATP-dependent Ca^{2+} pumping, and perhaps also via the Na^+-Ca^{2+} exchanger.[7] Ca^{2+} translocation via voltage-sensitive Ca^{2+} channels is a well-established route of entry of extracellular Ca^{2+} into excitable cells and perhaps many other cell types.[8] Moreover, it is clear now that activation of such channels can be finely controlled by intracellular messenger-mediated phosphorylation events.[8,9] In addition, many have considered that Ca^{2+} entry across the plasma membrane may be *directly* mediated by activation of channels distinct from voltage-sensitive Ca^{2+} channels.[7-9] The existence and characterization of such channels has not been conclusively described; however, it seems clear that at least the prolonged responses to many Ca^{2+}-coupled receptors are dependent on external Ca^{2+} and may involve entry of Ca^{2+} across the plasma membrane,[10] as discussed later.

In addition to the plasma membrane, internal organelles also play an important role in the maintenance of cytosolic Ca^{2+}. Mitochondria are known to actively accumulate Ca^{2+}[11] via a process dependent on a membrane potential existing across the internal membrane. However, from most observations it appears that mitochondria can only accumulate Ca^{2+} when free Ca^{2+} levels are high, that is, at or above 1 to 10 μM; thus it is unlikely that they contribute significantly in either the maintenance of physiological cytosolic Ca^{2+} levels or the induction of Ca^{2+}-signaling events within cells. In contrast, it appears certain that other Ca^{2+}-accumulating organelles within cells are active in both respects. Thus, endoplasmic reticulum (ER) in a variety of cell types has been observed to sequester large quantities of Ca^{2+}.[12-15] Using permeabilized nonmuscle cells, it is clear from a number of different studies that nonmitochondrial organelle(s) exist which accumulate Ca^{2+} via high affinity (ATP + Mg^{2+})-dependent Ca^{2+} pumping activity (see, for example, References 16 and 17). Such internal Ca^{1+} pumps are analogous in function to those of the plasma membrane, however, a number of features distinguish the internal and plasma membrane pumping activities.[17] Interestingly, these distinguishing characteristics are remarkably consistent with those features which serve to distinguish sarcolemmal and sarcoplasmic reticulum (SR) Ca^{2+} pumps in muscle tissues.[17] Thus, it has been suggested that ER in nonmuscle cells may fulfill at least some of the specialized Ca^{2+} regulatory functions ascribed to SR in muscle. However, although analogies exist with respect to Ca^{2+} accumulation, it is becoming increasingly apparent that the Ca^{2+} release mechanisms of SR and ER are quite distinct. It should also be noted that whereas the SR is a structurally identifiable organelle with a clearly defined Ca^{2+}-regulatory function, the role of ER in Ca^{2+} signaling within nonmuscle tissues is considerably more tenuous; thus, the involvement of ER in Ca^{2+} mobilizing events is concluded from indirect evidence with no proven localization as yet of these mechanisms to this specific organelle. Indeed, recent evidence presented by Krause suggests that Ca^{2+}-accumulating organelles which are *distinct* from ER may be involved in Ca^{2+}-regulatory responses in cells;[18] these organelles have been termed "calcisomes". In spite of the imprecise identity of Ca^{2+}-releasing organelles, ER is frequently referred to as being the organelle from which Ca^{2+} release occurs in response to inositol phosphates, the actions of which are discussed next.

B. INOSITOL PHOSPHATE-ACTIVATED CALCIUM MOVEMENTS

The recent advances in understanding both the metabolism and action of the inositol phosphates derived from receptor-mediated phospholipase C activation have had an enormous impact on the Ca^{2+}-signaling field.[2-5] It is currently held that an important direct product of phosphoinositide breakdown is inositol 1,4,5-trisphosphate (together with its 1,2-cyclic derivative), and that this molecule has proven effectiveness in releasing intracellular Ca^{2+} in a large variety of cells. The metabolism of this product is complex. Although it is not the purpose of this chapter to relate formation and breakdown of each of the products (reviewed in References 4 and 5), brief mention of the major derivatives is given here since certain of these may also have roles in modifying Ca^{2+} movements in cells. 1,4,5-IP_3 undergoes either phosphorylation or dephosphorylation. 5'-phosphatase activity in cells cleaves IP_3 to the largely inactive 1,4-IP_2 product. Alternatively, 3'-kinase activity can phosphorylate IP_3 to produce inositol 1,3,4,5-tetrakisphosphate (IP_4), which is itself a substrate for the 5'-phosphatase producing in this case inositol 1,3,4-trisphosphate. Whereas the latter molecule has little, if any, Ca^{2+}-releasing activity relative to 1,4,5-IP_3, the IP_4 molecule has been reported to exert indirect effects on Ca^{2+} mobilization.[19-21] Thus, IP_4 was observed by Irvine and Moor[19,20] to induce Ca^{2+} mediated effects in oocytes; these effects appear to be dependent on the presence of IP_3 and also to be induced only in the presence of external Ca^{2+}. Interpretation of the results may imply that IP_4 induces the entry of Ca^{2+} into the IP_3-releasable pool perhaps from outside the cell.[19,20,22] In a recent report, Petersen and co-workers[21] describe a similar synergism between the effects of Ip_3 and IP_4 on activation of K^+ channels in lacrimal gland; similar conclusions on the possible permissive effect of IP_4 on the action of IP_3-mediated Ca^{2+} mobilization were presented. More direct synergistic effects of IP_3 and IP_4 on Ca^{2+} have been reported by Spät et al.[23] Thus, it was observed that the extent of IP_3-mediated Ca^{2+} release from liver microsomal membrane vesicles was significantly increased in the presence of IP_4. At present, although it seems likely that IP_4 does exert effects, it is unclear whether it may directly control Ca^{2+} fluxes, whether it modifies the IP_3-induced release process, or whether it has indirect effects through alternation of the metabolism of IP_3, for example, by competing with IP_3 at the 5'-phosphatase level.

The action of IP_3 on the release of Ca^{2+} from what is believed to be ER occurs via a process that resembles activation of a channel.[24] This conclusion has been drawn from a number of observations including the remarkably temperature-insensitive activation of Ca^{2+} release in response to IP_3.[25,26] Direct electrophysiological evidence for an IP_3-activated channel has not yet been published; however, promising results have been discussed and more definitive studies are expected. Studies using labeled IP_3 have identified a binding site for IP_3 within cells, with kinetics and specificity similar to that for activation for Ca^{2+} release.[27-29] The isolation and apparent purification of an IP_3-binding protein which may be the IP_3 receptor has recently been reported.[30] Thus, whereas studies on the molecular structure and mechanism of the site of action of IP_3 are in their infancy, it is likely that much will come to light in the near future.

II. GUANINE NUCLEOTIDE-ACTIVATED CALCIUM RELEASE

In the last 3 years some important information has come to light on a guanine nucleotide-activated process that appears to directly effect profound and rapid movements of Ca^{2+} within a wide variety of cells. The following sections describe the nature of the effects of GTP on Ca^{2+} movements, their relationship to the actions of IP_3, and finally, the possible mechanism of activation of GTP-induced Ca^{2+} translocation. In this section we will consider the characteristics of the fluxes of Ca^{2+} activated by GTP.

A. ORIGINAL OBSERVATIONS

The first observation on the action of IP_3 in inducing Ca^{2+} release were made using permeabilized cell systems of several different types.[31,32] Some difficulty was encountered in observing the effects of IP_3 on isolated microsomal membrane vesicle fractions.[33] Such difficulties probably reflected either the lability of the IP_3-activated release process under lengthy vesicle purification procedures, and/or a low yield of intact vesicles derived from the IP_3-sensitive intracellular organelle. Dawson and colleagues were approaching this problem using liver microsomes in which they had observed small effects of IP_3.[33] In attempting to augment this response, Dawson observed that GTP enhanced the effectiveness of IP_3, and that this effect was promoted by polyethylene glycol.[34] Undertaking similar experiments with microsomes isolated from cultured N1E-115 neuroblastoma cells, we observed a rather different response.[35] With these microsomes, addition of IP_3 effected release of a small fraction (approximately 10%) of releasable Ca^{2+}. When both GTP and IP_3 were added simultaneously, a much larger release of Ca^{2+} was observed. However, in contrast to the results of Dawson, it was observed that GTP alone was highly effective in releasing Ca^{2+}.[35] The effect of GTP was rapid and profound, more than 50% of total accumulated Ca^{2+} being released from the microsomal membrane vesicles within a few seconds. As described below, the nucleotide specificity and sensitivity of the GTP effect were remarkable. The high GTP-sensitivity was considered possible since during their isolation the microsomes had undergone considerable washing and hence were largely devoid of endogenous nucleotides. With this in mind, it was reasoned that the permeabilized cell preparations used extensively in prior Ca^{2+} flux analyses,[17] having been subjected to less washing procedures, would be a less suitable preparation on which to observe GTP-induced Ca^{2+} fluxes. However, this prediction was incorrect, and in fact the permeabilized cell preparations became the system of choice on which most of the characteristics of GTP-activated Ca^{2+} movements were determined. The rapidity and extent of Ca^{2+} release activated by GTP in permeabilized N1E-115 cells is shown in Figure 1A where it is apparent that 50% of accumulated Ca^{2+} has been released within a period of only 30 s.[36] The effect is almost as rapid as adding the ionophore A23187, although the extent of release is not so complete as the latter, an observation that suggests heterogeneity of Ca^{2+}-accumulating compartments (see below).

B. GUANINE NUCLEOTIDE SENSITIVITY AND SPECIFICITY

The guanine nucleotide-activated release of Ca^{2+} observed using either permeabilized cells[36] or microsomes derived from cells[35] has remarkably high sensitivity to GTP. As shown in Figure 1B, the K_m for GTP measured in permeabilized N1E-115 cells is 0.75 μM. The effect also has very considerable nucleotide specificity. Release is not observed with GMP, cyclic GMP, (either $2'3'$ or $3'5'$), nor also with the nonhydrolyzable analogs of GTP, GTPγS, or GppNHp (see Figure 2). The latter is an important observation since it suggests a divergence in guanine nucleotide specificity from that of the known G proteins which are in fact super-stimulated by nonhydrolyzable analogs. Other nucleoside triphosphates including ITP, UTP, and CTP have no effect on Ca^{2+} movements (Figure 1A); these nucleotides are still largely ineffective when added at concentrations up to 1 mM. Submicromolar GTP concen-

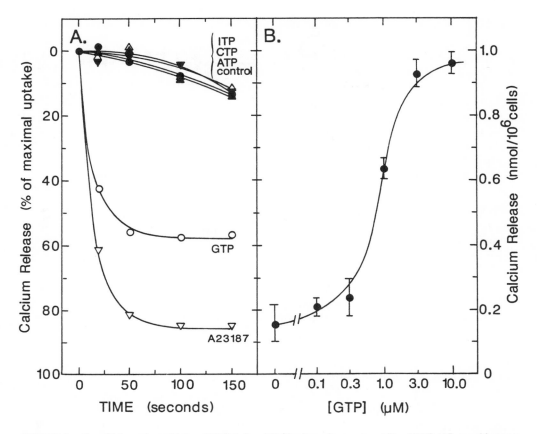

FIGURE 1. Specificity and sensitivity of GTP-induced Ca^{2+} release from permeabilized N1E-115 neuroblastoma cells. (A) Cells were loaded for 4 minutes with labeled Ca^{2+} under "cytosolic-like" conditions (140 mM KC1, 10mM NaC1, 2.5 mM MgCl$_2$, 0.1 μM free Ca^{2+}, 1 mM ATP, Hepes-KOH pH 7.0) at which time the following additions were made: 10 μM GTP (○), 10 μM ITP (▲), 10 μM CTP (△), 10 μM ATP (▼), 5 μM A23187 (▽), or control medium (●). Release was terminated at the times shown by La^{3+}-quenching and rapid filtration to determine the amount of Ca^{2+} remaining in the permeabilized cells. (B) Cells were loaded for 5 min followed by addition of the indicated GTP concentrations for 4 minutes after which remaining Ca^{2+} in the cells was determined as above; values are the means ± SD of six determinations. Details of the experimental procedures are given in Reference 36.

trations function to release Ca^{2+} in the presence of millimolar ATP concentrations (required to maintain constant Ca^{2+} pumping activity), indicating the exceptional specificity of the GTP-activated release process. It was observed that GDP does induce Ca^{2+} release, but only after a significant lag of about 30 s (Figure 2A); thereafter it releases Ca^{2+} to approximately the same extent as GTP. Results clearly indicate that this effect results from conversion of GDP to GTP via nucleoside diphosphokinase activity.[35,36] Thus, the effect of GDP is blocked by ADP (Figure 2B) which effectively competes for the nucleoside diphosphate site on NDPK.[37] In fact, GDP itself does not induce Ca^{2+} release; thus, GDPβS (which is not easily phosphorylated to GTPβS by NDPK) has no effect on Ca^{2+} release (Figure 2B). Moreover, not only is GDP without Ca^{2+}-releasing effects of its own, but it actually blocks the action of GTP, as shown in Figure 2C; (note that at 100 μM, GDP saturates NDPK activity and remains present for a longer period of time to compete with GTP). Further experimentation (in the presence of high ADP to prevent conversion of GDP to GTP) revealed that the inhibitory effect of GDP was competitive with respect to GTP with a K_i of approximately 3 μM;[36] GTPγS also blocks the effect of GTP, but rather surprisingly, GppNHp does not (Figure 2C). This differential inhibitory action of the non-

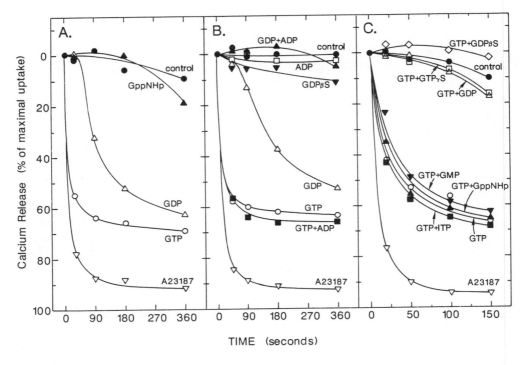

FIGURE 2. Influence of different guanine nucleotides on the release of Ca^{2+} from permeabilized N1E-115 cells. Loading of Ca^{2+} was for 4 min under standard conditions before additions. Conditions for the loading and release measurements were as described in Figure 1 and Reference 36. Release was determined after addition of either. (A) control buffer (●), 10 μM GTP (○), 20 μM GppNHp (▲), 20 μM GDP (△), or 5 μM A23187 (▽) (B) buffer control (●), 10 μM GTP (○), 10 μM GDP (△), 1 mM ADP (□), 10 μM GTP with 1 mM ADP (■), 10 μM GDP with 1 mM ADP (▲), 10 μM GDPβS (▼), or 5 μM A23187 (▽); (C) buffer control (●), 3 μM GTP (○), 3 μM GTP with 100 μM GppNHp (▲), 3 μM GTP with 100 μM GDP (△), 3 μM GTP with 100 μM GMP (▼), 3 μM GTP with 100 μM ITP (■), 3 μM GTP with 100 μM GTPγS (□), 3 μM GTP with 100 μM GDPβS (◇), 5 μM A23187 (▽). The addition of each of these agents or combinations of agents as shown were all made at zero-time.

hydrolyzable analogs has been a useful criterion for defining the specificity of the GTP-activated process and is referred to again later. The lack of direct action of GTPγS and its inhibitory effect on the action of GTP are evidence that GTP hydrolysis is required for the activation of Ca^{2+} release. In fact, a very slow release activated by GTPγS[26] may be consistent with slow cleavage of the phosphorothioate residue which is known to occur.[38] Further evidence for a GTP hydrolytic process being involved in activation of Ca^{2+} release derives from the competitive effect of GDP which indicates that either GTP or GDP can bind to the same site; presumably, the inhibitory effect of GDP arises through prevention of GDP dissociation after hydrolysis of GTP at the Ca^{2+} release-activating site.

C. CELLULAR AND SUBCELLULAR SPECIFICITY

It was important to establish whether the effectiveness of GTP in directly inducing Ca^{2+} release was an anomaly, perhaps restricted to the N1E-115 neuronal cell line used in our early studies. Experiments undertaken on a quite unrelated cell type, the DDT_1MF-2 smooth muscle cell line derived from hamster vas deferens,[39] suggest this is not the case. Thus, a sensitive, specific, and substantial GTP-dependent release of Ca^{2+} was observed using permeabilized DDT_1MF-2 cells loaded with Ca^{2+}, with pronounced effectiveness of as low as 0.1 μM GTP in the presence of 1 mM ATP.[40] In addition to the DDT_1MF-2 cell line, we have measured almost identical effects of GTP on Ca^{2+} release using permeabilized cells from the rat BC_3H-1 smooth muscle cell line, and from the human WI-38 normal embryonic lung fibroblast cell line. Using microsomal membrane vesicle fractions prepared from DDT_1MF-2 cells by methods similar to those described for N1E-115 cell-derived microsomes,[35] we have observed GTP effects on Ca^{2+} release almost identical to those seen with permeabilized cells. Furthermore, using microsomes derived from guinea pig parotid gland, Henne and Söling[41] have observed very similar effects on release of accumulated Ca^{2+} induced by GTP. The observations of Jean and Klee[42] on GTP- and IP_3-mediated Ca^{2+} release from microsomes derived from NG108-15 neuroblastoma $\times$ glioma hybrid cells are also very consistent with our findings.

The process of Ca^{2+} release is specific to a nonmitochondrial Ca^{2+}-sequestering organelle, believed to be ER; no effects of guanine nucleotides of IP_3 can be observed on Ca^{2+} fluxes across mitochondrial or plasma membranes.[35,40] The observation that less than 100% of Ca^{2+} release from ER is effected by GTP or IP_3 suggests that only a subcompartment of ER contains the activatable efflux mechanisms. Although we have no direct proof that ER is a source of GTP-releasable Ca^{2+}, interpretation of the effects of oxalate (described later), a known permeator of the ER membrane,[40,43,44] may indicate that ER is indeed a site of action of both GTP and IP_3. Moreover, we now know that GTP indeed modifies the movements of Ca^{2+} associated with the IP_3-releasable Ca^{2+} pool and hence that GTP and IP_3 can act on the same Ca^{2+} pool, as described below.

D. MECHANISM OF CALCIUM TRANSLOCATION
1. Reversibility of Action of GTP

Determination of the nature of the Ca^{2+} translocation process activated by GTP has been a major area of investigation. Either of two distinct possibilities appeared likely: first, GTP could activate a channel process to permit the flow of Ca^{2+} out of the organelle(s) into which Ca^{2+} is sequestered; alternatively, GTP could activate a fusion between organelle membranes and result in the release or transfer of Ca^{2+}. In the latter case, it would be very unlikely that such a process would be reversible, that is, that the two fused membranes could be returned to the unfused state with the same original enclosed volume. Recently, we reported that GDP at least partially reverses the prior effectiveness of GTP suggesting some degree of reversibility of the action of GTP.[36] Since then, a more definitive indication of the reversibility of the effect of GTP has come from a simpler study involving washing of cells after GTP activation.[40] Thus, it has been observed that after activation of the GTP-dependent Ca^{2+} release process (with up to 100 μM GTP), the effectiveness of GTP can be substantially (more than 70%) reversed by simple washing of the GTP-treated permeabilized cells with GTP-free medium. In such experiments, cells that had been treated with GTP under conditions that activate Ca^{2+} release were thoroughly washed; after this treatment Ca^{2+} uptake proceeded to an extent approaching that of untreated cells, that is, the ability of ER to accumulate Ca^{2+} was largely restored. Moreover, such GTP-pretreated, washed cells respond again to a further application of GTP, indicating that the release process can be reactivated by GTP. It would be difficult to reconcile this reversibility with a membrane fusion process activated by GTP; in other words, the effects of a direct membrane fusion event would be unlikely to be reversed by washing and result in the restoration of almost normal Ca^{2+} retention, as observed. It should be noted, however, that structural and biophysical measurements undertaken by Dawson and co-workers suggest that fusion of membranes *can* follow GTP treatment of microsomal vesicles.[45,46] At present this question is unresolved.

2. Membrane Interactions

Recent electron microscopic analysis of membrane vesicles treated with GTP has suggested that the action of GTP, while not necessarily involving membrane fusion, may be

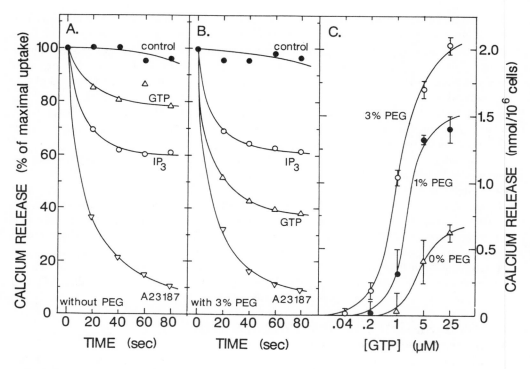

FIGURE 3.   Effects of polyethylene glycol (PEG) on Ca^{2+} release from permeabilized N1E-115 cells induced by either GTP or IP_3. In (A) and (B), the time-dependence of Ca^{2+} release is shown after addition of either 5 μM IP_3 (lc), 10 μM GTP (△), 5 μM A23187 (▽), or buffer control (●), each added at zero time to the Ca^{2+}-loaded cells. Uptake and release of Ca^{2+} in (A) were undertaken in the absence of PEG. In (B), 3% w/v PEG (M_r = 6,000) was present during both the 6-min Ca^{2+} uptake period and during Ca^{2+} release. In (C), the effect of different PEG concentrations (% w/v) on the GTP-dependency of Ca^{2+} release was observed; Ca^{2+} release was for 2 min after addition of the GTP concentrations shown, and was undertaken in the presence of either 3% PEG (○), 1% PEG (●), or in the absence of PEG (△). The data in (C) are means ± SD of triplicate measurements. Further details of the experimental procedures are given in Reference 26.

promoted by close association between membranes. It is now well established that the effects of GTP on Ca^{2+} release are promoted by 1 to 3% polythylene glycol.[26,35,36] Thus, as shown in Figure 3, while in the absence of PEG, GTP induces a significant release of Ca^{2+}, this effect is substantially increased in the presence of 3% PEG (Figure 3B). The effect of PEG is to increase both the sensitivity to and maximal release induced by GTP (Figure 3C). Although PEG is a known fusogen when present above 25% w/v,[47] we believe that the effect of PEG in enhancing Ca^{2+} release is unlikely to involve membrane fusion. Thus, our recent studies have analyzed by electron microscopy the appearance of isolated microsomal membrane vesicles derived from N1E-115 cells after GTP-treatment with or without PEG. As shown in Figure 4, GTP was without any effect on vesicle appearance, whereas 3% PEG induces a very clear coalescence of vesicles into tightly associated conglomerates with very few free or unattached vesicles. The effect of PEG was not visibly altered by GTP. It may, therefore, be concluded that GTP itself does not induce any observable alteration in vesicle structure or association. However, the striking effectiveness of PEG is good evidence to suggest that the effect of GTP in inducing Ca^{2+} movements is promoted by a condition that increases the close association between membranes. This may be an important clue to the action of GTP, as discussed in detail below. Thus, we consider that close association between membranes might be sufficient to permit the GTP-induced event which could involve formation of some type of junctional process between membranes perhaps permitting the flow of Ca^{2+}, thereafter, it is possible that under certain conditions membrane fusion may occur.

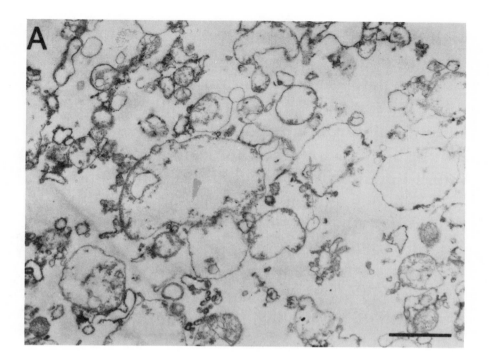

FIGURE 4.   Electron microscopic analysis of thin sections of microsomal membrane vesicles derived from permeabilized N1E-115 cells after treatment with PEG and/or GTP. Microsomal membrane vesicles were isolated from permeabilized N1E-115 cells as described previously.[35] Vesicles were subjected to the same Ca^{2+} uptake conditions as in Figure 1, with the exception that labeled Ca^{2+} was omitted and PEG was not included in samples A and C (3% PEG was present in samples B and D). In samples C and D, 10 μM GTP was also included. After 10 min exposure to uptake conditions, vesicle samples were prepared for and examined by electron microscopy, as described in Reference 40. Bars in each micrograph represent 1 μm.

III. RELATIONSHIP BETWEEN IP$_3$- AND GTP-ACTIVATED CALCIUM RELEASE

A. DISTINCTIONS IN MECHANISM OF ACTION OF IP$_3$ AND GTP

A number of distinctions exist between the actions of IP$_3$ and GTP on Ca^{2+} release, as described in our recent report.[26] First, IP$_3$-mediated release is unaffected by either GDP or GTPγS, both of which block the action of GTP on Ca^{2+} release, as described above. Second, PEG which considerably promotes GTP-activated release (as shown in Figure 3) does not alter the action of IP$_3$; the lack of effect of PEG on IP$_3$-induced Ca^{2+} release (see Figure 3A and B) suggests that IP$_3$ functions via a mechanism that does not require close membrane interactions. A third distinction between the actions of IP$_3$ and GTP is the temperature-dependency of their effects. Thus, the effect of IP$_3$ is remarkably insensitive to temperature changes, the rate of IP$_3$-induced Ca^{2+} release being reduced by only 20% when the temperature is decreased from 37 to 4°C; this contrasts with the complete abolition of the effectiveness of GTP at the lower temperature.[26] The latter result is consistent with GTP activating release via an enzymic mechanism perhaps involving GTP hydrolysis, whereas the action of IP$_3$ is unlikely to involve an enzymic step. A fourth major distinction between the actions of IP$_3$ and GTP concerns their Ca^{2+} dependency. Thus, IP$_3$-induced Ca^{2+} release, in contrast to that induced by GTP, is modified by the free Ca^{2+} concentration, as shown in Figure 5. In this experiment, Ca^{2+} uptake and release were measured at free Ca^{2+}

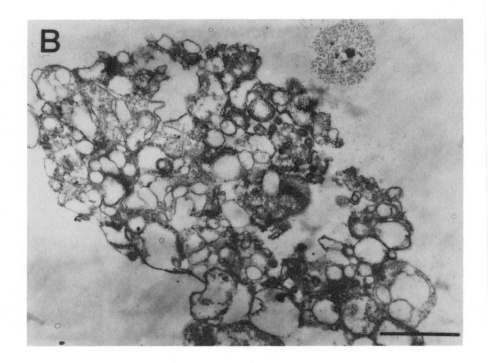

FIGURE 4B.

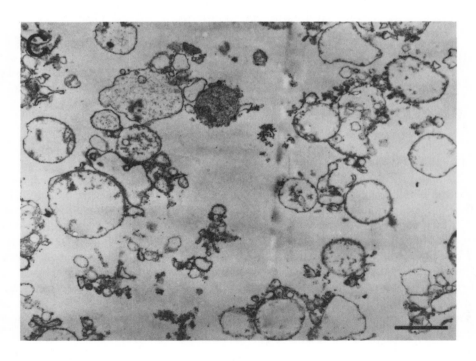

FIGURE 4C.

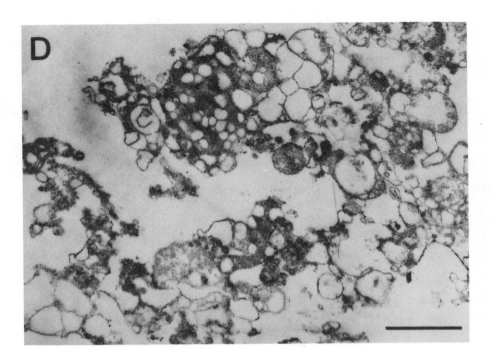

FIGURE 4D.

concentrations of either 0.1 μM (the concentration normally used in experiments), or 10 μM. The effect of IP_3 is reduced by 50% at 1 μM free Ca^{2+} and completely abolished at 10 μM Ca^{2+}. In contrast, GTP induces identical fractional Ca^{2+} release over this entire range of free Ca^{2+}. The inhibition of IP_3-mediated Ca^{2+} release over this free Ca^{2+} range is a significant observation indicating that the IP_3 release process is under negative feedback control from the level of Ca^{2+}, a potentially important regulatory response.[26] Interestingly, recent work from Worley et al.[29] indicates that binding of labeled IP_3 to its putative membrane receptor has almost identical Ca^{2+} sensitivity, suggesting that the feedback effect may exist at the IP_3 binding step.

From the above results we would conclude that the activation of Ca^{2+} release by GTP or IP_3 occurs via distinct mechanisms. Several of these distinctions between the actions of GTP and IP_3 have also been reported by Henne and Söling[41] using either liver- or parotid-derived microsomes, and by Jean and Klee[42] using microsomes derived from NG108-15 neuroblastoma $\times$ glioma hybrid cells. It is concluded that the relative temperature insensitivity and rapidity of IP_3-induced Ca^{2+} release are consistent with its probable direct activation of a Ca^{2+} channel, a conclusion in agreement with the observations of others.[24,25] In contrast, GTP appears to effect release by a temperature-sensitive process which probably involves the enzymic hydrolysis of the terminal phosphate from GTP.

B. IP_3- AND GTP-RELEASABLE CALCIUM POOLS

The IP_3- and GTP-induced Ca^{2+} release processes both function on a similar intracellular Ca^{2+}-sequestering compartment. Yet, the size of the releasable pools of Ca^{2+} are distinct, in the N1E-115 cell line, for example, that induced by GTP being approximately twice the size of the IP_3-releasable pool. Thus, as shown in Figure 6 using permeabilized N1E-115 cells, following maximal Ca^{2+} release by GTP, IP_3 is not effective in releasing further Ca^{2+} (Figure 6B); however, following maximal release by IP_3 (approximately 30% of accumulated Ca^{2+}), GTP does effect a further release of Ca^{2+} (Figure 6A), in fact, down to the level

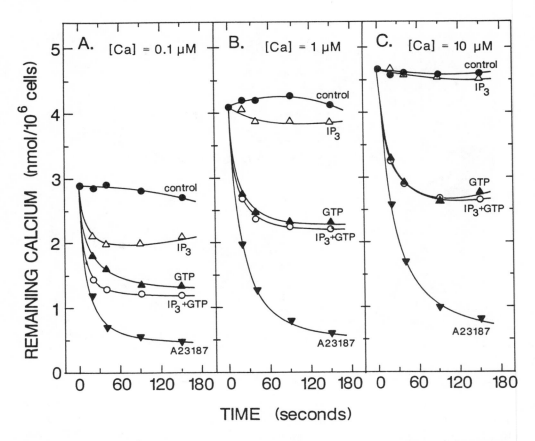

FIGURE 5.   Free Ca^{2+} concentration-dependence of IP_3- and GTP-activated Ca^{2+} release from permeabilized N1E-115 cells. Both uptake and release were undertaken in the presence of 10 μM $CaCl_2$ either buffered with EGTA to 0.1 μM (A) or 1 μM free Ca^{2+} (B), or unbuffered to nominally give 10 μM free Ca^{2+} (C). Release of Ca^{2+} was measured for the times indicated after the addition of either 10 μM IP_3 (△), 10 μM GTP (▲), 10 μM IP_3 together with 10 μM GTP (○), 5 μM A23187 (▼), or control buffer (●). Uptake and release were undertaken in the presence of 5 μM ruthenium red to suppress mitochondrial Ca^{2+} accumulation. See Reference 26 for further details of experimental procedures.

GTP could induce when added alone (that is, approximately 60% of accumulated Ca^{2+}). These results suggest that three compartments exist; one sensitive to both GTP and IP_3, another releasing only in response to GTP, and a third not releasing in response to either agent. Thus, it is apparent that although the GTP-releasable pool differs from the IP_3-releasable pool in being larger, at least a significant proportion of accumulated Ca^{2+} lies within a pool which can be released by either of the two agents. In other words, it appears that all of the Ca^{2+} within the IP_3-sensitive Ca^{2+} also exists within the GTP-activated Ca^{2+} pool. This implies a probable proximal relationship between the IP_3- and GTP-activated Ca^{2+} release processes, and permits us to consider the existence of possible coupling events linking their modes of action.

IV. MECHANISM OF ACTION OF GTP ON CALCIUM MOVEMENTS

Before attempting to infer the mechanism of activation of Ca^{2+} release induced by GTP, another important observation must be considered. We recently observed that GTP can induce an entirely opposite effect on Ca^{2+} movements in the presence of oxalate, that is,

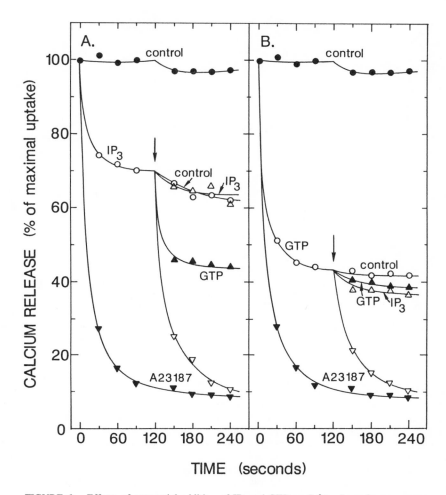

FIGURE 6. Effects of sequential addition of IP$_3$ and GTP on Ca^{2+} release from permeabilized N1E-115 neuroblastoma cells. Ca^{2+} release was measured after loading for 5 min in the presence of 0.1 μM free Ca^{2+}, under the standard conditions (see Figure 1). (A) Immediately following uptake, release was observed after addition of either 10 μM IP$_3$ (○), 5 μM A23187 (▼), or buffer control (●); after 120 s of release in the presence of IP$_3$, measurement of release was continued after further additions of either 10 μM IP$_3$ (△), 10 μM GTP (▲), 5 μM A23187 (▽), or buffer control (○). (B) Immediately following uptake, release was observed after addition of either 10 μM GTP (○), 5 μM A23187 (▼), of buffer control (●); after 120 s of release in the presence of GTP, release was continued after further addition of either 10 μM IP$_3$ (△), 10 μM GTP (▲), 5μM A23187 (▽), or buffer control (○). In each case, samples of the Ca^{2+}-loaded permeabilized cell suspension were removed followed by rapid filtration and washing as described in Reference 40.

GTP induces *uptake* as opposed to *release* of Ca^{2+} when oxalate is present. Although this observation appeared at first anomalous, it has in fact provided an important piece of evidence in formulating a model for the action of GTP.

A. GTP-ACTIVATED CALCIUM UPTAKE
1. Oxalate Effects

The evidence provided above suggested that a simple GTP-mediated membrane fusion event was not entirely consistent with the observed release of Ca^{2+} induced by GTP. However, in order to further approach this question, experiments were designed to determine whether oxalate-precipitation of Ca^{2+} would prevent release activated by GTP. As shown

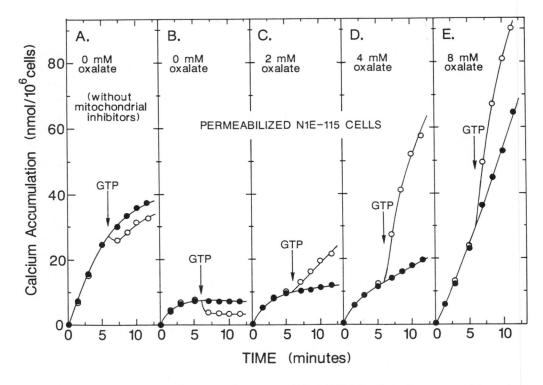

FIGURE 7.   GTP-activated movements of Ca^{2+} in permeabilized N1E-115 cells in the presence of increasing concentrations of oxalate. Experimental conditions were as described in References 40 and 43. These were essentially the same as in Figure 1 with the exception that EGTA was absent and total $CaCl_2$ was 30 M. Experiments were undertaken either in the absence of mitochondrial inhibitors (A), or in the presence of 5 μM ruthenium red and 10 μM oligomycin (B — E). K-oxalate was either absent (A and B), or present from the beginning of uptake at a final concentration of 2 mM(C), 4 mM(D), or 8 mM(E). After 6 min of uptake, 10 μM GTP (○) or control buffer (●) was added to the permeabilized cell suspensions. Aliquots from the incubation vials were taken at the times shown, and Ca^{2+} remaining within cells was determined by rapid La^{3+}-quenching and filtration as described in Reference 40.

in our previous studies[17] and well established in many different cell types,[12-15] the ER is permeable to anions including oxalate and phosphate which can diffuse into the ER lumen and hence promote a large increment in Ca^{2+} uptake due to formation of insoluble Ca^{2+} oxalate or phosphate complexes. In order to further investigate how GTP activates Ca^{2+} release, we tested to see if oxalate-precipitated Ca^{2+} within ER could be released by GTP, thus a negative result would again militate against a simple membrane fusion event accounting for release and would instead argue in favor of a more selective channel mechanism, passage of precipitated Ca^{2+} through which would not be expected. However, as shown in Figure 7, an unexpected and profound increase in Ca^{2+} uptake was observed in the presence of oxalate, a remarkable and entirely opposite effect to that observed in the absence of oxalate. The effect is observed with concentrations of oxalate (2 mM) which show very little effect on uptake of Ca^{2+} in the absence of GTP (Figure 7C). When oxalate is present at a concentration inducing linear uptake of Ca^{2+} (Figure 7E), GTP still activates an additional increase in the rate of uptake. This phenomenon is not restricted to particular cell types, thus an identical effect of GTP on Ca^{2+} uptake in the presence of oxalate was observed using either N1E-115 neuroblastoma or DDT_1MF-2 smooth muscle cells.

2. Identity of GTP-Activated Calcium Uptake and Release

In view of the paradoxically opposite effects of GTP in the presence or absence of oxalate, it was important to establish whether both GTP-mediated events resulted from a common mechanism of action of GTP. It is now clear from a large number of observations that this is the case. For example, as shown in Figure 8, the GTP-dependence of Ca^{2+} uptake-induced in the presence of oxalate is almost identical to that of the release induced without oxalate (Figure 1). In the experiment depicted in Figure 8, only the GTP-activated uptake component is shown. From linearization of these data, a K_m for GTP of 0.9 μM is obtained, very close to the value of 0.75 μM derived from Ca^{2+} release data described above and reported in earlier studies.[36] Further studies have revealed that the uptake of Ca^{2+} induced by GTP in the presence of oxalate is promoted by PEG in a manner very similar to GTP-activated Ca^{2+} release without oxalate.[43] Thus, although both GTP-activated release and uptake are observable in the absence of PEG, both effects are considerably augmented in the presence of 3% PEG.

In addition to these similarities between Ca^{2+} uptake and release induced by GTP, the nucleotide specificity profiles of the two processes closely coincide.[43] A particularly important observation in this regard is that both effects of GTP show the same differential specificity towards the actions of nonhydrolyzable GTP analogs. Thus, GTP-activated Ca^{2+} uptake in the presence of oxalate is activated by neither GTPγS nor GppNHp; however, GTPγS, but not GppNHp, completely blocks the action of GTP. As described above, it is assumed that the site at which nucleotides bind distinguishes between the two GTP analogs. The noneffectiveness of these analogs in promoting GTP-like effects is in clear distinction to the effects of guanine nucleotides on known G-protein activities (such as those modulating adenylate cyclase) where nonhydrolyzable analogs are maximally or super-effective. The identity between the GTP-activated release and uptake processes is further exemplified by observing the effects of GDP on Ca^{2+} movements. Thus, GDP gives a full but delayed uptake response[43] which exactly coincides with its effect on release in the absence of oxalate[36] (see Figure 2). Moreover, as with release, GDP-mediated uptake is blocked by ADP indicating its action arises from conversion to GTP via nucleoside diphosphokinase activity. In the presence of ADP, GDP blocks the action of GTP; GDPβS, which does not activate uptake, also blocks the action of GTP.

These data reveal almost exact correlation between parameters affecting GTP-activated uptake and release. A summary of these effects is given in Table 1. Such data provide very strong evidence suggesting that the same GTP-activated process mediates both uptake and release of Ca^{2+} in the presence and absence of oxalate, respectively. The only divergence between the two processes is the effectiveness of vanadate which blocks GTP-induced uptake but does not block GTP-activated release.[43] However, as discussed below, the proposed model for the actions of GTP accounts for this difference.

B. CLUES TO THE ACTION OF GTP IN MEDIATING CALCIUM MOVEMENTS

Several important pieces of information are derived from the above data which together are consistent with a model invoking a GTP-mediated conveyance of Ca^{2+} across membranes and perhaps between organelles. Before discussing this conclusion, let us summarize this new information. First, we have observed that a discrete pool of GTP-releasable Ca^{2+} exists in cells, a pool that may incorporate within it a smaller IP_3-releasable Ca^{2+} pool; yet, despite the overlap between pools, we have provided substantial evidence suggesting distinctions between the mechanisms of GTP and IP_3 in *activating* Ca^{2+} release.[26] Second, since PEG promotes both the effects of GTP and a clearly observable membrane coalescence at the same PEG concentrations (1 to 3%),[40] it is probable that activation of Ca^{2+} movements within cells is related to the occurrence of close appositions between membranes. Third, although direct fusion between membranes could account for some of the effects of GTP on Ca^{2+} movements, the observed reversibility of the effects of GTP together with the nonreleasability of oxalate-complexed Ca^{2+} by GTP, would argue against a simple GTP-

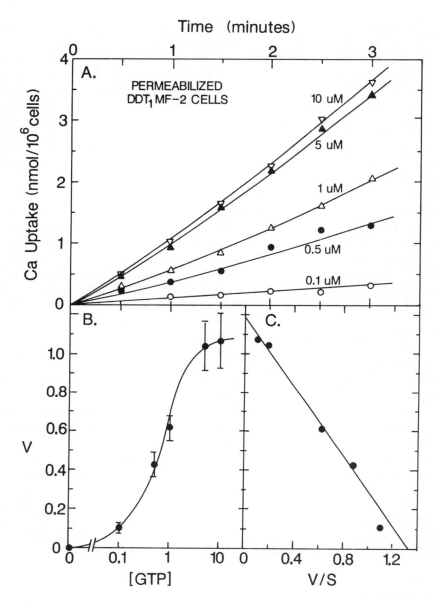

FIGURE 8.   The GTP-dependence of GTP-activated Ca^{2+} uptake in permeabilized DDT_1MF-2 cells. Cells were incubated in uptake medium under the conditions described in Figure 7 (see Reference 43), with 10 μM oligomycin, 4 mM K-oxalate, and 3% PEG present in all incubations. Ca^{2+} uptake into cells proceeded without any GTP addition for 6 min, at which time GTP was added to give final GTP concentrations of 0.1 μM (○), 0.5 μM (●), 1 μM (△), 5 μM (▲), or 10 μM (▽). GTP-dependent uptake of Ca^{2+} (that is, uptake above that observed over the same time period without GTP addition) is plotted against time after GTP addition in panel A. Mean values ± SD calculated by linear regression of the rate of GTP-activated Ca^{2+} uptake (in units of nmol $Ca^{2+}/10^6$ cells/min) over the 3-min uptake period are plotted against GTP concentration in panel B. Eadie-Hofstee analysis of the same data is shown in panel C from which a K_m value for GTP of 0.9 μM is obtained.

TABLE 1

**Summary of Parameters of GTP-activated Calcium Release and Calcium
Uptake in the Absence and Presence of Oxalate, Respectively**

Parameter or condition	Calcium release (observed without oxalate)	Calcium uptake (observed with oxalate)
K_m for GTP	0.75 μM	0.9 μM
10 μM GDP	Delayed full effect	Delayed full effect
10 μM GDP (+ 1 mM ADP)	No effect	No effect
100 μM GDPβS	No effect	No effect
10 μM GTP + 100 μM GDP (+ 1 mM ADP)	GTP effect blocked	GTP effect blocked
10 μM GTP + 100 μM GDPβS	GTP effect blocked	GTP effect blocked
10 μM GTPγS	Slight effect	Slight effect
10 μM GTP + 100 μM GTPγS	GTP effect blocked	GTP effect blocked
10 μM GppNHp	No effect	No effect
10 μM GTP + high 100 μM GppNHp	GTP effect not blocked	GTP effect not blocked
1—3% PEG	Stimulated	Stimulated
1 mM Vanadate	No effect	Blocked

Note: Each of the parameters of GTP-activated Ca^{2+} uptake observed in the presence of oxalate refers to data presented in References 40 and 43. The observations relating to Ca^{2+} release (in the absence of oxalate) were published in prior reports (References 26, 35, and 36). Explanations and details of the conditions described are given in the text.

mediated fusion event between membrane surfaces as being the direct cause of Ca^{2+} movements. Fourth, there seems little doubt that the proces of GTP-activated Ca^{2+} uptake in the presence of oxalate occurs via a mechanism probably identical to that by which GTP activates release of Ca^{2+}, in spite of the apparent opposite nature of these two GTP-mediated events.

This last piece of information appeared the most perplexing, yet ironically it may provide the most significant clue to the action of GTP. Thus, it is likely that oxalate promotes the uptake of Ca^{2+} into a discrete Ca^{2+}-accumulating pool. It is well known that the ER membrane is permeable to anions including oxalate and phosphate; hence passive entry of oxalate permits the formation of clearly observable insoluble complexes within the lumen of ER in cells;[12-15] the entry of such anions may be mediated via a nonselective anion transporter activity analogous to that functioning in the SR membrane of muscle.[48] It is also apparent from our previous studies that, whereas Ca^{2+} accumulation in permeabilized cells and isolated microsomal membrane vesicles is oxalate-promoted, the accumulation of Ca^{2+} within purified inverted plasma membrane vesicles via the high affinity plasma membrane Ca^{2+} pump is not enhanced by oxalate.[17] Since we have shown that these plasma membrane vesicles can indeed accumulate high intravesicular Ca^{2+} concentrations,[49-51] more than sufficient to be precipitated in the presence of millimolar oxalate concentrations, we conclude that such membranes are largely impermeable to oxalate or phosphate. Thus, there is a good precedent for the existence of membranes through which passage of oxalate does not occur.

C. THE GTP-ACTIVATED TRANSMEMBRANCE CALCIUM "CONVEYANCE" MODEL

With the knowledge that distinct membranes exist which are differentially permeable to oxalate, we propose that in the presence of oxalate, GTP promotes uptake of Ca^{2+} as the result of a GTP-mediated movement of Ca^{2+} from a nonoxalate-permeable pool which actively pumps Ca^{2+}, to another Ca^{2+}-pumping pool which is freely permeable to oxalate. Thus, it is envisaged that GTP promotes a transmembrane conveyance of Ca^{2+} between such pools by activating some type of junctional process between the two membranes (see

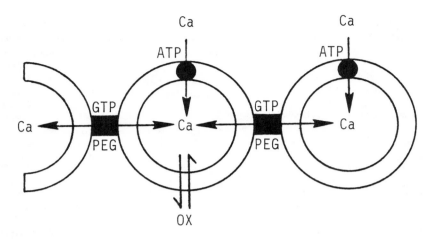

FIGURE 9.   Hypothetical model explaining the two effects of GTP on Ca^{2+} movements in cells, that is, GTP-mediated Ca^{2+} release and GTP-mediated Ca^{2+} uptake in the absence and presence of oxalate, respectively. Details of the evidence and an explanation of the proposed "conveyance" of Ca^{2+} between open and closed compartments via a transmembrane Ca^{2+} translocation process are given in the text and in References 43 and 44.

Figure 9). Alternative schemes involving GTP-promoted oxalate-permeability or enhanced Ca^{2+} pumping are possible; but why then should an almost identical GTP-dependent process mediate movement (release) of Ca^{2+} in the absence of oxalate? In the model depicted in Figure 9, the oxalate-permeable pool is very likely to be the ER or a subcompartment thereof; the nature of the putative nonoxalate permeable pool is uncertain. Although the plasma membrane has been rendered permeable in our studies, it is possible that separate enclosed membranes derived from the plasma membrane might exist within the cell; such autonomous vesicles would be largely protected from the permeabilizing effects of saponin. The postulated process of junction formation between membranes would obviously be promoted by conditions which favor close appositions between membranes, as occurs in the presence of PEG. The action of GTP is envisaged as a necessary factor in either inducing the formation of junctions or activating the movement of Ca^{2+} ions through junctional processes arising by random of PEG-promoted membrane interactions. Such transfer of Ca^{2+} would be activated by terminal phosphate hydrolysis from GTP; when GTP is washed away, then the continued operation of such transfer would be terminated as indicated by the reversibility experiments described above.[40]

Based on this model, how could GTP-mediated Ca^{2+} release be accounted for? It seems entirely possible that the same type of junctional connections could be formed between intact organelles such as ER and nonclosed membranes, perhaps the plasma membrane. In this case, transmembrane conveyance of Ca^{2+} would result in release of Ca^{2+} to the medium (see Figure 9). If such a conveyance of Ca^{2+} to the outside could be mediated by GTP how could GTP induce a buildup of Ca^{2+} within the oxalate-permeable pool? If the hypothetical GTP-activated junctional processes transmit only small solutes between pools (as gap junctions between cells are known to do), then a precipitate of the Ca^{2+}-oxalate complex would not be expected to be transferred. Thus, in the experiments described above, oxalate and Ca^{2+} have been permitted to equilibrate within the oxalate permeable pool; addition of GTP may result in a substantial "injection" of Ca^{2+} from a nonoxalate permeable pool; this Ca^{2+} would be immediately precipitated due to the excess oxalate present. When GTP and oxalate are added simultaneously at the beginning of uptake, GTP causes a prolonged inhibition of Ca^{2+} uptake due to activation of the release process. With time and in the presence of sufficient oxalate, there is a gradual increase in uptake followed eventually by a sustained uptake which proceeds at a rate approaching the maximal rate of uptake observed when GTP is added after oxalate (see below). The initial phase of this type of response is presumably due to the continued release of Ca^{2+} to the exterior thus preventing sufficient buildup of Ca^{2+} to that critical level at which precipitation with oxalate occurs. Irrespective of when oxalate is added, the Ca^{2+}-conveyance model predicts that Ca^{2+}-pumping activity is essential to sustain GTP-activated Ca^{2+} uptake in the presence of oxalate, a prediction clearly confirmed by the blocking action of vanadate.

A significant question that has dominated the physiological implications of the GTP-activated Ca^{2+} movements we have described is how the high levels of GTP within cells (0.1 to 0.3 mM) can be reconciled with the extreme sensitivity of the GTP-activated process; thus, it was argued that under physiological conditions, the intracellular pool acted upon by GTP wold be permanently depleted.[52] However, by implicating a transfer of Ca^{2+} only between actively pumping organelles (and possibly with the outside of the cell), there would not be any collapse of existing gradients. Thus, the release that is observed with GTP may only reflect an artificially imposed, diminished external Ca^{2+} level that is a consequence of using permeabilized cells. In other words, such release could actually represent reversed movement of Ca^{2+} through a system that normally exists to convey Ca^{2+} perhaps to replenish the intracellular pool. The implication here is that such interpool communication may normally exist between organelles in intact cells but be reversed when cells are broken and GTP washed away. Alternatively, the functioning of such Ca^{2+} communication between organelles may be regulated *in situ* by another cytosolic factor.

D. GTP-INDUCED LOADING OF THE IP₃-RELEASABLE POOL

With the above model in mind, perhaps the most relevant question to be addressed was the relationship between the pools of Ca^{2+} modified by GTP and that Ca^{2+} pool sensitive to IP_3. This area of investigation has produced some important results. A first determination to be made was to ascertain whether IP_3 releases Ca^{2+} from an oxalate-permeable or -impermeable pool. This question is largely answered by the data shown in Figure 10. Using permeabilized cells from the DDT_1MF-2 smooth muscle and the N1E-115 neuroblastoma cell lines, IP_3 in the absence of oxalate reduces Ca^{2+} uptake by 50 and 30%, respectively, (Figure 10A and C), effects entirely consistent with the extent of Ca^{2+} release observed following IP_3 addition to Ca^{2+}-loaded cells, as described above. In the presence of oxalate a sustained increase in the rate of ATP-dependent Ca^{2+} accumulation is observed (Figure 10B and D) consistent with formation of the insoluble Ca-oxalate complex and hence a reduced rate of Ca^{2+} efflux.[17,43] Importantly, 10 μM IP_3 (a maximally effective concentration) completely eliminates the increment in Ca^{2+} uptake induced by oxalate in permeabilized DDT_1MF-2 cells (Figure 10B) indicating that IP_3 activates Ca^{2+} release from an oxalate-permeable pool. Although not completely abolishing oxalate-enhanced Ca^{2+} uptake, the effectiveness of IP_3 is very similar using permeabilized N1E-115 cells (Figure 10D); hence in these cells, whereas IP_3 does release from an oxalate-permeable pool, a small fraction of this pool may be unresponsive to IP_3. As stated above, it is well established that the ER membrane is permeable to anions including oxalate hence permitting clearly observable precipitation of Ca^{2+} within the ER lumen when oxalate is presented intracellularly.[12-15] Thus, these data, while not providing definitive proof, are consistent with the view that the source of IP_3-mobilizable Ca^{2+} is the ER or at least a subcompartment thereof.

Although, as described above, there are clear distinctions between the mechanisms by which IP_3 and GTP activate Ca^{2+} movements, the data shown in Figure 11 clearly establish a link between the actions of the two effectors. When added from the start of uptake, GTP and IP_3 inhibit the accumulation of Ca^{2+} in a nonadditive manner (Figure 11A) consistent with the extent of release described above. In the presence of oxalate, the action of GTP is very different from that of IP_3 (Figure 11B). Thus, whereas IP_3 merely inhibits accumulation,

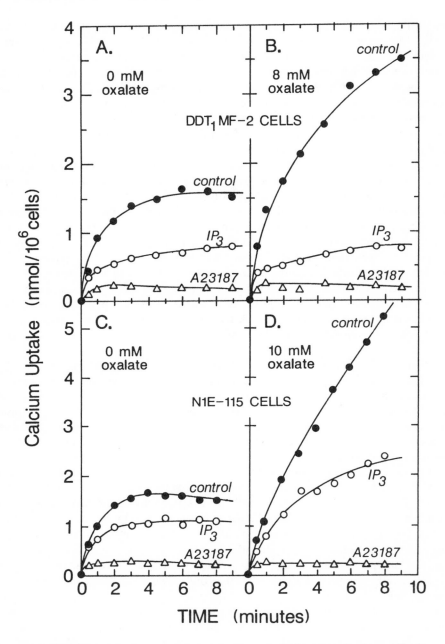

FIGURE 10. IP$_3$-mediated inhibition of oxalate-dependent Ca^{2+} uptake into permeabilized DDT$_1$MF-2 smooth muscle cells (A and B) or N1E-115 neuroblastoma cells (C and D). Uptake of Ca^{2+} was started at zero-time by addition of ATP and labeled Ca^{2+} to gently stirring cells in uptake medium and was terminated by rapid filtration of aliquots of cells removed at the indicated times, as described for Figure 1. Oxalate was either absent from the uptake medium (A and C) or was present from the start of uptake at either 8 mM (B) or 10 mM (D). Uptake was measured either under standard conditions (●) or in the presence of 10 μM IP$_3$ (○) or 5 μM A23187 (△) added to incubation vials in each case at zero-time. The medium contained 50 μM CaCl$_2$ buffered to 0.1 μM with EGTA; further experimental details are given in Reference 44.

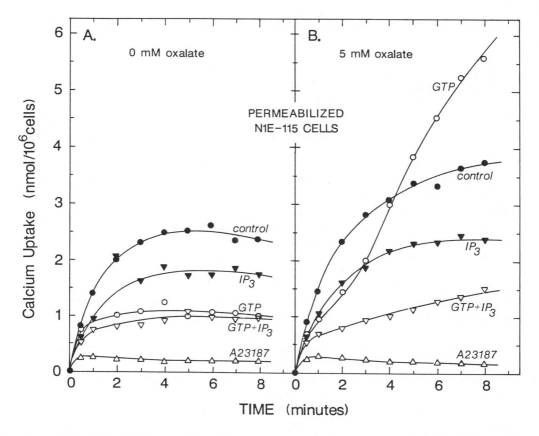

FIGURE 11. IP$_3$-induced reversal of the GTP-activated Ca^{2+} uptake phase in permeabilized N1E-115 neuroblastoma cells. ATP-dependent uptake of Ca^{2+} was measured at the indicated times after addition of ATP and labeled Ca^{2+} to cells as described in Figure 10 (see Reference 44). Incubations were conducted either in the absence of oxalate (A) or in the presence of 5 mM oxalate (B). Uptake proceeded under otherwise standard conditions (●), or after addition of either 10 μM GTP (○), 10 μM IP$_3$ (▼), 10 μM GTP together with 10 μM IP$_3$ (▽), or 5 μM A23187 (△). Additions of oxalate, IP$_3$, GTP, and A23187 were all made at zero-time

GTP shows a biphasic effect. This effect of GTP is interpreted to further support the model in Figure 9 since it shows that the two opposing GTP-activated movements of Ca^{2+} directly compete for access of Ca^{2+} to a common compartment. Thus, although initially release occurs resulting from interactions between closed and open compartments, thereafter, as the threshold of accumulated Ca^{2+} reaches that precipitable by oxalate, release of complexed Ca^{2+} is prevented and Ca^{2+} continues to accumulate at a higher rate reflecting the combined pumping activity of intact pools, Ca^{2+} movement between which has been activated by GTP. Most importantly, the GTP-induced enhanced Ca^{2+} uptake phase is almost completely abolished when IP$_3$ is added together with GTP indicating that IP$_3$ releases Ca^{2+} from the same pool into which GTP activates Ca^{2+} accumulation. These results obtained using permeabilized N1E-115 neuroblastoma cells have been repeated almost identically using permeabilized DDT$_1$MF-2 smooth muscle cells. It should be noted that IP$_3$ does not block the effects of GTP per se, since Ca^{2+} accumulation is reduced to a level well below that induced by IP$_3$; thus, it may be inferred that while IP$_3$ prevents the additional accumulation of Ca^{2+} activated by GTP it in fact permits the Ca^{2+}-releasing effects of GTP to dominate. These results provide direct evidence for the operation of both GTP- and IP$_3$-activatable Ca^{2+} transport mechanisms on the same pool of Ca^{2+}. Most significantly, they suggest that loading of Ca^{2+} within the IP$_3$-sensitive pool may be controlled by the GTP-activated Ca^{2+} translocation process.

V. CONCLUSIONS AND MODEL FOR THE ACTIONS OF IP₃ AND GTP

The proposed scheme of GTP-activated Ca^{2+} movements accounts for all the observed effects of GTP and oxalate on Ca^{2+} movements. Alternative schemes invoking direct effects of GTP on Ca^{2+} pumping or GTP-enhanced movements of oxalate are inherently unlikely since they do not account for rapid GTP-mediated Ca^{2+} release. In fact GTP-mediated Ca^{2+} release occurs in the presence of vanadate and in the absence of ATP.[43] Also, in a recent report, Hamachi et al.[53] described similar GTP-enhanced uptake of Ca^{2+} in the presence of oxalate; although no explanation was offered for the effect, direct experiments revealed no effect of GTP on oxalate movements. While recent work from Dawson and colleagues[45,46] suggests membrane fusion may account for the effects of GTP, as stated above, the observations we have made on reversibility of the effects of GTP and on electron microscopic analysis of the structure microsomal membrane vesicles treated with GTP, together argue against a simple membrane fusion process being activated by GTP.[7] Although GTP-hydrolysis is clearly implicated in the process of GTP-activated Ca^{2+} translocation,[26,36] it is presently unclear whether terminal phosphate is transferred to water (as in the case of a GTPase reaction), or whether a kinase-mediated mechanism transfers phosphate to another substrate molecule. Evidence for the former was recently presented by Nicchitta et al.,[54] whereas a GTP-induced protein phosphorylation possibly associated with Ca^{2+} release was claimed by Dawson et al.[55]

Based on several important conclusions drawn from the data given in Figures 10 and 11, the scheme described above to account for the effects of GTP can be extended to encompass the action of IP₃. First, from the data in Figure 11A and data described earlier, it is apparent that the IP₃-releasable Ca^{2+} pool is both smaller than and contained within the GTP-activatable pool. Second, based on the results shown in Figure 10, the pool from which IP₃ induces release is itself permeable to oxalate. Third, and most significant, this IP₃-releasable Ca^{2+} pool is indeed the same pool that can be loaded with Ca^{2+} via the GTP-induced Ca^{2+}-translocating process, as shown in Figure 11B. These observations suggest to us that the IP₃-releasable Ca^{2+} pool *is* the oxalate-permeable subcompartment of the GTP-activatable pool, as depicted in the model shown in Figure 12. Thus, we assume that the efficient operation of the IP₃-activated Ca^{2+} channel enhances efflux of Ca^{2+} from this pool effectively enough to prevent sufficient buildup of Ca^{2+} to reach the oxalate-precipitable threshold. Interestingly, when the experiment shown in Figure 11B is conducted with 10 m*M* oxalate or higher (data not shown), some GTP-dependent buildup of Ca^{2+} does occur at later times in the presence of IP₃, suggesting that by lowering the oxalate-threshold, even the rapid release effected by IP₃ is insufficient to prevent a significant buildup of Ca^{2+}.

The direct reversal of the effect of GTP by IP₃ provides a strong argument for considering that indeed both IP₃ and GTP can act upon a common pool of Ca^{2+}. Such conclusions are reminiscent of our earlier "flux reversal" studies which provided direct proof for the co-existence of specific plasma membrane Ca^{2+} and Na^+ flux mechanisms in a single population of synaptic membrane vesicles.[49,50] The most significant implication of the scheme shown in Figure 12 is that a close interrelationship likely exists between the actions of IP₃ and GTP. We had previously speculated that this might be the case[43] but had no proof for such a scheme. The data presented in Figures 10 and 11 provide for the first time direct evidence that both IP₃ and GTP can modify the same compartment of Ca^{2+} in spite of their likely distinct mechanisms of action.

It is very possible that the GTP-regulated Ca^{2+}-translocating process may control the size of the IP₃-induced Ca^{2+} signal by permitting IP₃ to release Ca^{2+} from a more extensive internal Ca^{2+} pool. Moreover, the same process may regulate the loading/or replenishment of Ca^{2+} within the IP₃-releasable pool. Such potential regulation derives much relevance

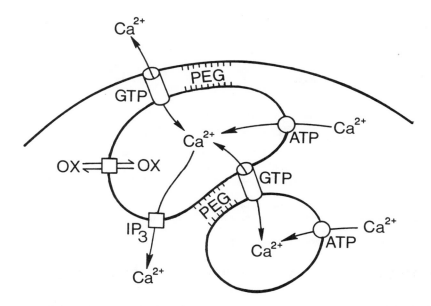

FIGURE 12.   Proposed scheme for the movements of Ca^{2+} induced by GTP and IP_3. The model proposes that separate ATP-dependent Ca^{2+} pumping compartments exist which are distinct with respect to both IP_3-releasability and oxalate permeability, and that GTP mediates Ca^{2+} translocation between such compartments perhaps via activation of intermembrane junctional processes. It is further proposed that GTP-activated Ca^{2+} release occurs via the same mechanism except involving interactions between the surfaces of closed and nonclosed membranes. As described previously (References 26, 35, 36, 40 and 43), the effects of GTP on both uptake and release of Ca^{2+} are promoted by low concentrations of PEG (1 to 3%) which also promote the formation of close appositions between membrane surfaces (References 40 and 43); however, PEG is not essential and the same GTP-activated Ca^{2+} movements are still observable without PEG (References 26 and 43). Details of the evidence indicating the site of action of IP_3 and the proposed mechanism of GTP in this scheme are described in the text and References 40, 43, and 44.

from the considerable recent attention that has been directed towards the possible mechanisms by which the IP_3-releasable Ca^{2+} pool may be replenished from the outside. Thus, Putney[10] has suggested that external Ca^{2+} entry may be directed into this pool and hence account for the frequently observed prolonged responses to receptor-induced signals which are dependent on extracellular Ca^{2+}. Recently, Irvine and Moor[19,20] have presented experimental evidence suggesting the possible involvement of inositol 1,3,4,5-tetrakisphosphate (IP_4) in inducing Ca^{2+} entry; in fact, their studies on activation of sea urchin eggs are consistent with the possibility that IP_4 may promote entry of external Ca^{2+} into the IP_3-releasable pool via a mechanism[20] remarkably similar to the scheme described here for the movements of Ca^{2+} induced by GTP. We are currently investigating whether this putative action of IP_4 is related to GTP-activated Ca^{2+} movements and/or whether IP_4 may modulate GTP-induced Ca^{2+} translocation.

REFERENCES

1. **Hokin, M. R. and Hokin, L. E.,** Enzyme secretion and the incorporation of ^{32}P into phospholipides of pancreas slices, *J. Biol. Chem.,* 203, 967, 1953.
2. **Berridge, M. J. and Irvine, R. F.,** Inositol trisphosphate, a novel second messenger in cellular signal transduction, *Nature*, 312, 315, 1984.

3. **Gill, D. L.**, Receptors coupled to calcium mobilization, *Adv. Cyclic Nucleotide Protein Phos. Res.*, 19, 195, 1985.

4. **Majerus, P. W., Connolly, T. M., Deckmyn, H., Ross, T. S., Bross, T. E., Ishii, H., Bansal, V. S., and Wilson, D. B.**, The metabolism of phosphoinositide-derived messenger molecules, *Science*, 234, 1519, 1986.

5. **Berridge, M. J.**, Inositol trisphosphate and diacylglycerol: two interacting second messengers, *Annu. Rev. Biochem.*, 56, 159, 1987.

6. **Carafoli, E.**, Intracellular calcium homeostasis, *Annu. Rev. Biochem.*, 56, 395, 1987.

7. **Gill, D. L.**, Receptor-mediated modulation of plasma membrane calcium transport, *Horiz. Biochem. Biophys.*, 6, 199, 1982.

8. **Miller, R. J.**, Multiple calcium channels and neuronal function, *Science*, 235, 46, 1987.

9. **Tsien, R. W., Bean, B. P., Hess, P., Lansman, J. B., Nilius, B., and Nowycky, M. C.**, Mechanisms of calcium channel modulation by β-adrenergic agents and dihydropyridine calcium agonists, *J. Mol. Cell. Cardiol.*, 18, 691, 1986.

10. **Putney, J. W.**, A model for receptor-regulated calcium entry, *Cell Calcium*, 7, 1, 1986.

11. **Hansford, R. G.**, Relation between mitochondrial calcium transport and control of energy metabolism, *Rev. Physiol. Biochem. Pharmacol.*, 102, 1, 1985.

12. **Henkart, M. P., Reese, T. S., and Brinley, F. J.**, Endoplasmic reticulum sequesters calcium in the squid giant axon, *Science*, 202, 1300, 1978.

13. **McGraw, C. F., Somlyo, A. V., and Blaustein, M. P.**, Localization of calcium in presynaptic nerve terminals, *J. Cell Biol.*, 85, 228, 1980.

14. **Wakasugi, H., Kimura, T., Haase, W., Kribben, A., Kaufmann, R., and Schulz, I.**, Calcium uptake into acini from rat pancreas: evidence for intracellular ATP-dependent calcium sequestration, *J. Memb. Biol.*, 65, 205, 1982.

15. **Burton, P. R. and Laveri, L. A.**, The distribution, relationships to other organelles, and calcium-sequestering ability of smooth endoplasmic reticulum in frog olfactory axons, *J. Neurosci.*, 5, 3047, 1985.

16. **Burgess, G. M., McKinney, J. S., Fabiato, A., Leslie, B. A., and Putney, J. W., Jr.**, Calcium pools in saponin-permeabilized guinea pig hepatocytes, *J. Biol. Chem.*, 258, 15336, 1983.

17. **Gill, D. L. and Chueh, S. H.**, An intracellular (ATP + Mg^{2+})-dependent calcium pump within the N1E-115 neuronal cell line, *J. Biol. Chem.*, 260, 9289, 1985.

18. **Krause, K.-H.**, Calsiosomes: evidence for a new type of organelle regulating intracellular Ca^{2+} in phagocytes, *Proc. 7th Int. Washington Spring Symp. (Cell Calcium Metabolism)*, Washington D.C.

19. **Irvine, R. F. and Moor, R. M.**, Micro-injection of inositol 1,3,4,5-tetrakisphosphate activates sea urchin eggs by a mechanism dependent on external Ca^{2+}, *Biochem. J.*, 240, 917, 1986.

20. **Irvine, R. F. and Moor, R. M.**, Inositol(1,3,4,5)tetrakisphosphate-induced activation of sea urchin eggs requires the presence of inositol trisphosphate, *Biochem. Biophys. Res. Commun.*, 146, 284, 1987.

21. **Morris, A. P., Gallacher, D. V., Irvine, R. F., and Petersen, O. H.**, Synergism of inositol trisphosphate and inositol tetrakisphosphate in activating Ca^{2+}-dependent K^+ channels, *Nature*, 330, 653, 1987.

22. **Michell, R. H.**, A second messenger function for inositol tetrakisphosphate, *Nature*, 324, 613, 1986.

23. **Spät, A., Lukács, G. L., Eberhardt, I., Kiesel, L., and Runnebaum, B.**, Binding of inositol phosphates and induction of Ca^{2+} release from pituitary microsomal fractions, *Biochem. J.*, 244, 493, 1987.

24. **Muallem, S., Schoeffield, M., Pandol, S., and Sachs, G.**, Inositol trisphosphate modification of ion transport in rough endoplasmic reticulum, *Proc. Natl. Acad. Sci. U.S.A.*, 82, 4433, 1985.

25. **Smith, J. B., Smith, L., and Higgins, B. L.**, Temperature and nucleotide dependence of calcium release by myo-inositol 1,4,5-trisphosphate in cultured vascular smooth muscle cells, *J. Biol. Chem.*, 260, 14413, 1985.

26. **Chueh, S. H. and Gill, D. L.**, Inositol 1,4,5-trisphosphate and guanine nucleotides activate calcium release from endoplasmic reticulum via distinct mechanisms, *J. Biol. Chem.*, 162, 13883, 1986.

27. **Baukal, A. J., Guillemette, G., Rubin, R. P., Spät, A., and Catt, K. J.**, Binding sites for inositol trisphosphate in the bovine adrenal cortex, *Biochem. Biophys. Res. Commun.*, 133, 532, 1985.

28. **Spät, A., Bradford, P. G., McKinney, J. S., Rubin, R. P., Putney, J. W., Jr.**, A saturable receptor for ^{32}P-inositol-1,4,5-trisphosphate in hepatocytes and neutrophils, *Nature*, 319, 514, 1986.

29. **Worley, P. F., Baraban, J. M., Supattapone, S., Wilson, V. S., and Snyder, S. H.**, Characterization of inositol trisphosphate receptor binding in brain, *J. Biol. Chem.*, 262, 12132, 1987.

30. **Supattapone, S., Worley, P. F., Baraban, J. M., and Snyder, S. H.**, Solubilization, purification, and characterization of an inositol trisphosphate receptor, *J. Biol. Chem.*, 263, 1530, 1988.

31. **Streb, H., Irvine, R. F., Berridge, M. J., and Schulz, I.**, Release of Ca^{2+} from a non-mitochondrial intracellular store in pancreatic acinar cells by inositol 1,4,5-trisphosphate, *Nature*, 306, 67, 1983.

32. **Burgess, G. M., Godfrey, P. P., McKinney, J. S., Berridge, M. J., Irvine, R. F., and Putney, J. W., Jr.**, The second messenger linking receptor activation to internal Ca^{2+} release in liver, *Nature*, 309, 63, 1984.

33. **Dawson, A. P. and Irvine, R. F.,** Inositol(1,4,5)trisphosphate-promoted Ca^{2+} release from microsomal fractions of rat liver, *Biochem. Biophys. Res. Commun.,* 120, 858, 1984.

34. **Dawson, A. P.,** GTP enhances inositol trisphosphate-stimulated Ca^{2+} release from rat liver microsomes, *FEBS Lett.,* 185, 147, 1985.

35. **Ueda, T., Chueh, S. H., Noel, M. W., and Gill, D. L.,** Influence of inositol 1,4,5-trisphosphate and guanine nucleotides on intracellular calcium release within the N1E-115 neuronal cell line, *J. Biol. Chem.,* 261, 3184, 1986.

36. **Gill, D. L., Ueda, T., Chueh, S. H., and Noel, M. W.,** Ca^{2+} release from endoplasmic reticulum is mediated by a guanine nucleotide regulatory mechanism, *Nature,* 320, 461, 1986.

37. **Kimura, N. and Shimada, N.,** GDP does not mediate but rather inhibits hormonal signal to adenylate cyclase, *J. Biol. Chem.,* 258, 2278, 1983.

38. **Eckstein, F.,** Nucleoside phosphorothioates, *Annu. Rev. Biochem.,* 54, 367, 1985.

39. **Norris, J. S., Gorski, J., and Kohler, P. O.,** Androgen receptors in a Syrian hamster ductus deferens tumour cell line, *Nature,* 248, 422, 1974.

40. **Chueh, S. H., Mullaney, J. M., Ghosh, T. K., Zachary, A. L., and Gill, D. L.,** GTP and inositol 1,4,5-trisphosphate-activated intracellular calcium movements in neuronal and smooth muscle cell lines, *J. Biol. Chem.,* 262, 13857, 1987.

41. **Henne, V. and Söling, H-D.,** Guanosine 5'-triphosphate releases calcium from rat liver and guinea pig parotid gland endoplasmic reticulum independently of inositol 1,4,5-trisphosphate, *FEBS Lett.,* 202, 267, 1986.

42. **Jean, B. and Klee, C. B.,** Calcium modulation of inositol 1,4,5-trisphosphate-induced calcium release from neuroblastoma × glioma hybrid (NG108-15) microsomes, *J. Biol. Chem.,* 261, 16414, 1986.

43. **Mullaney, J. M., Chueh, S. H., Ghosh, T. K., and Gill, D. L.,** Intracellular calcium uptake activated by GTP; evidence for a possible guanine nucleotide-induced transmembrane conveyance of intracellular calcium, *J. Biol. Chem.,* 262, 13865, 1987.

44. **Mullaney, J. M., Yu, M., Ghosh, T. K., and Gill, D. L.,** Calcium entry into the inositol 1,4,5-triphosphate-releasable calcium pool is mediated by a GTP-regulatory mechanism, *Proc. Natl. Acad. Sci. U.S.A.,* 85, 2499, 1988.

45. **Dawson, A. P., Hills, G., and Comerford, J. G.,** The mechanism of action of GTP on Ca^{2+} efflux from rat liver microsomal vesicles, *Biochem. J.,* 244, 87, 1987.

46. **Comerford, J. G. and Dawson, A. P.,** The mechanism of action of GTP on Ca^{2+} efflux from rat liver microsomal vesicles; measurement of vesicle fusion by fluorescence energy transfer, *Biochem. J.,* 249, 89, 1988.

47. **Hui, S. W., Isac, T., Boni, L. T., and Sen, A.,** Action of polyethylene glycol on the fusion of human erythrocyte membranes, *J. Membr. Biol.,* 84, 137, 1985.

48. **Martonosi, A. N.,** Transport of calcium by sarcoplasmic reticulum, in *Calcium in Cell Function,* Vol. 3, Cheung, W. Y., Ed., Academic Press, New York, 1985, 37.

49. **Gill, D. L., Grollman, E. F., and Kohn, L. D.,** Calcium transport mechanisms in membrane vesicles from guinea pig brain synaptosomes, *J. Biol. Chem.,* 256, 184, 1981.

50. **Gill, D. L.,** Sodium channel, sodium pump, and sodium-calcium exchange activities in synaptosomal plasma membrane vesicles, *J. Biol. Chem.,* 257, 10986, 1982.

51. **Gill, D. L., Chueh, S. H., and Whitlow, C. L.,** Functional importance of the synaptic plasma membrane calcium pump and sodium-calcium exchanger, *J. Biol. Chem.,* 259, 10807, 1984.

52. **Baker, P. F.,** GTP and calcium release, *Nature,* 320, 395, 1986.

53. **Hamachi, T., Hirata, M., Kimura, Y., Ikebe, T., Ishimatsu, T., Yamaguchi, K., and Koga, T.,** Effect of guanosine triphosphate on the release and uptake of Ca^{2+} in saponin-permeabilized macrophages and the skeletal-muscle sarcoplasmic reticulum, *Biochem. J.,* 242, 253, 1987.

54. **Nicchitta, C. V., Joseph, S. K., and Williamson, J. R.,** Polyethylene glycol-stimulated microsomal GTP hydrolysis; relationship to GTP-mediated Ca^{2+} release, *FEBS Lett.,* 209, 243, 1986.

55. **Dawson, A. P., Comerford, J. G., and Fulton, D. V.,** The effect of GTP on inositol 1,4,5-trisphosphate-stimulated Ca^{2+} efflux from a rat liver microsomal fraction, *Biochem. J.,* 234, 311, 1986.

G Proteins in Sensory Transduction

Chapter 7

G PROTEINS IN OLFACTORY NEURONS

Richard C. Bruch

TABLE OF CONTENTS

I.   Introduction...124
      A.   Properties of Olfactory Neurons...124
      B.   Olfactory Receptor Characteristics......................................125

II.   G Proteins in Olfactory Transduction..127
      A.   G-Protein Identification and Localization...............................127
      B.   G-Protein Regulation of Olfactory Transduction.........................128
            1.   G-Protein Coupling to Olfactory Receptors.......................128
            2.   Adenylate Cyclase...128
            3.   Phosphoinositide Hydrolysis.....................................130

III. Conclusions...131

Acknowledgments..132

References...132

I. INTRODUCTION

For many years, the molecular mechanisms underlying stimulus activation of olfactory neurons and the biochemical events involved in signal transduction in these cells were poorly understood. It has been known from electrophysiological studies that olfactory stimuli elicit depolarization of the chemosensory membrane and that membrane depolarization ultimately leads to action potential generation and synaptic transmission.[1,2] However, until recently, definition of the sequence of molecular events between the initial interaction of stimuli with the chemoreceptive membrane and regulation of subsequent electrical events was speculative and poorly characterized. In contrast to other sensory systems such as audition and vision, advances in our understanding of olfaction at the molecular and cellular levels have been complicated by several factors that are unique to chemoreception. Historically, uncertainties regarding delivery of volatile stimuli, choice of appropriate and experimentally amenable animal models, and ambiguity associated with the cellular specificity of stimulus-evoked responses, have contributed to an incomplete, and often controversial, description of the molecular and cellular mechanisms of olfaction. In addition, unlike audition and vision, a quantitative spectrum does not exist to describe the wide variety of stimuli that are recognized by olfactory neurons. The recent substantial progress toward elucidation of the molecular mechanisms of signal transduction in other sensory systems, such as photoreceptor cells,[3] has facilitated recent efforts to characterize the biochemical events that mediate stimulus-response coupling in olfactory neurons. This chapter describes current concepts of the biochemistry of vertebrate olfaction, particularly with regard to the role of G proteins in regulation of receptor-mediated second messenger events and modulation of ion channel activity. The literature since about 1980 is emphasized, since at that time, many of the earlier biochemical aspects of olfaction were published.[4]

A. PROPERTIES OF OLFACTORY NEURONS

In vertebrates, the olfactory epithelium (mucosa) is the peripheral tissue specialized for the detection and discrimination of olfactory stimuli. The epithelium is composed of two morphologically distinct regions: the chemosensory neuroepithelium, proximal to the external environment, and a deeper lamina propria. The neuroepithelium is mitotically active throughout the life span of the organism and undergoes continuous cellular degeneration and replacement. At the cellular level, the neuroepithelium is heterogeneous and is composed of three major cell types: olfactory receptor neurons, sustentacular cells, and basal cells.[1,2,5] The receptor cells, specialized for odorant detection, are primary chemosensory neurons that are separated by the supporting sustentacular cells. Basal cells provide a pool of progenitor cells that differentiate and replace senescent receptor cells. Secretory products from glands in the lamina propria, together with sustentacular cell secretions, provide the mucus layer that covers the apical surface of the neuroepithelium. Stimulus access to, and clearance from, the chemosensory apical surface depends on the physical properties of both the stimulus and the mucus. Stimulus transport rates in the aqueous mucus depend on several factors, including stimulus partition and diffusion characteristics, uptake and metabolism of some stimuli, and stimulus interaction with components in the mucus.[1,5]

Olfactory receptor cells are bipolar neurons with a single dendrite projecting toward the external environment and a single, unbranched, and nonmyelinated axon projecting through the lamina propria. The axons collectively form the olfactory nerve (cranial nerve I) and synapse directly with the central nervous system at the glomeruli of the olfactory bulb.[1,2,5] The apical ends of the receptor cell dendrites terminate in specialized extensions of the dendritic membrane. These membrane extensions on most receptor cells are cilia, although a second receptor cell type terminates in microvilli. Microvillar receptor neurons are particularly prominent in fish,[6] but have also been identified in humans.[7] Since differential stimulus selectivity of the two receptor cell types has been reported in some fish species,[8] but not in others,[9] the functional significance of the two receptor cell types remains unresolved.

A persistent problem that continues to limit biochemical studies of olfactory stimulus-response coupling is the unusual difficulties associated with isolation of purified preparations of the receptor cells. Although olfactory neurons express several characteristic biochemical markers that aid in their identification,[1,5,10] fractionation of dissociated cell suspensions from the olfactory epithelium to obtain purified or cultured preparations of the receptor cells has not been extensively investigated. Single-cell electrophysiological techniques have recently been successfully applied to morphologically identified neurons in mixed cell suspensions.[1,11-13] However, dissociated receptor cells have often not exhibited consistent stimulus-evoked responses and often display time-dependent alteration of their distinctive morphology.[14,15] The limited attempts to obtain enriched receptor cell preparations by methods that exploit differences in cell size and density have been only partially successful due in large part to the intrinsic cellular heterogeneity of the neuroepithelium.[14]

Since purified or cultured olfactory neuron preparations are not yet routinely available, most biochemical studies of olfactory reception and signal transduction have focused on the unique dendritic cilia. The variety of evidence implicating these organelles in olfactory reception,[1,2,5,16] and their physical proximity to the external environment, point to the cilia as the site of the initial interaction of stimuli with the chemoreceptive membrane. The cilia are readily isolated by a generally applicable "calcium shock" procedure. Differential centrifugation and further purification on discontinuous sucrose gradients yields an isolated cilia preparation that represents about 2 to 4% of the total neuroepithelial protein.[17-20] Isolated cilia preparations have become generally accepted as appropriate membrane preparations for biochemical studies of the initial events of olfactory reception and signal transduction, and thus may be regarded as the olfactory equivalent of the retinal rod outer segment preparation. In cross-section, the cilia display the characteristic "9 $\times$ 2 + 2" arrangement of microtubules surrounded by plasma membrane. Tubulin, the major axonemal structural protein, is particularly abundant in the cilia. The ciliary membrane is readily solubilized in nonionic detergent, providing a convenient method for separating membrane components from the axonemes which remain insoluble.[21] A variety of glycoproteins have been identified in the ciliary membrane by lectin reactivity.[18-21] One of these, gp95, is an abundant transmembrane glycoprotein that is exclusively localized in olfactory cilia and appears to be widely distributed across vertebrate species.[20,22,23] Although it has been proposed as a candidate olfactory receptor protein since it exhibits several properties expected of a membrane-associated receptor.[21,23] The functional role of gp95 in chemoreception has not been experimentally confirmed.

B. OLFACTORY RECEPTOR CHARACTERISTICS

It is generally accepted that olfactory neuron activation is initiated by stimulus interaction with specific, membrane-associated receptor proteins in the cilia.[1,2,16] However, the receptor hypothesis has not been rigorously established and remains controversial,[24,25] since the appropriate molecular species have not been identified or isolated. Unlike many hormone and neurotransmitter receptors, for which chemically pure and biologically selective ligands are available, ligand binding studies with volatile olfactory stimuli are complicated by ligand purity and solubility, low binding affinity, and nonspecific association of hydrophobic compounds with biological membranes. Although odorant-binding sites have been identified in the olfactory epithelium of mammals and putative binding proteins have been isolated,[24,26] confirmatory evidence that the binding sites or isolated proteins correspond to receptor proteins has not been reported. Recently, a soluble pyrazine-binding protein was isolated from bovine olfactory epithelium.[27,28] The isolated homogeneous protein retained high affinity pyrazine-binding activity and it was therefore initially suggested to represent an olfactory receptor protein. Subsequent immunohistochemical analysis showed that the pyrazine-binding protein was found in the mucus and localized in the secretory glands.[29] The same or similar protein from frog exhibits partial sequence homology with serum transport proteins.[30] Although the pyrazine-binding protein exhibits odorant-binding activity and shares some sequence homology to other soluble binding proteins, its secretory nature and cellular localization are unexpected properties for an olfactory receptor. The role of this protein in chemoreception remains unresolved, although it has been suggested to be involved in odorant transport and solubilization in the mucus.[29]

Ligand binding studies with isolated membrane preparations from the olfactory epithelium of fish have provided the majority of the currently available evidence supporting the receptor hypothesis. The initial experiments of Cagan and co-workers showed that binding sites for amino acids, olfactory stimuli for many fish, were present in membranes derived from olfactory epithelium homogenates from trout.[31] These investigators also first described the successful isolation of olfactory cilia and further showed that the amino acid binding sites detected in crude membranes were retained in the isolated cilia.[16,17,32] These initial studies showed that stimulus binding in the isolated cilia was consistent with generally expected specificity, saturability, and reversibility criteria for a membrane-associated receptor. In addition, good agreement was obtained between the rank order of binding and electrophysiological potency, suggesting that the binding data represented the interaction of these stimuli with physiologically relevant receptors. The selectivity of the binding sites was consistent with the conclusion of a limited number of amino acid olfactory receptors that recognized structurally related ligands. Neutral L-amino acids, such as alanine, serine, and threonine, but not basic amino acids, competed for the same binding site, suggesting the existence of a discrete site (site TSA) for these stimuli. Similarly, basic amino acids, such as L-arginine and L-lysine, which interacted with a common binding site (site L), did not compete for site TSA. Brown and Hara obtained similar results in membranes derived from trout olfactory epithelium homogenates, but concluded that the binding data represented amino acid transport rather than ligand-receptor interaction.[33]

Recent ligand binding studies in salmon[34] and catfish[35,36] have also documented the existence of olfactory amino acid binding sites that exhibit similar affinity ($K_D = 10^{-7} - 10^{-5}$ M) and selectivity as the sites described in trout. In salmon, the binding data paralleled stimulus specificity and potency characteristics determined in behavioral assays.[34] In isolated cilia from catfish, amino acid binding generally paralleled the specificity, potency, and stereoselectivity properties expected for the binding sites from electrophysiological studies.[37,38] The likely glycoprotein nature of the amino acid receptors in isolated cilia from catfish was also suggested by lectin inhibition of receptor binding.[35] Three lectins differentially inhibited ligand binding to the receptors for L-arginine and L-alanine by a mechanism that decreased the number of binding sites with no change in affinity. Lectin inhibition of receptor binding for both ligands was completely reversible in the presence of the appropriate competing monosaccharides, indicating that the inhibition of receptor binding resulted from specific carbohydrate recognition. Although differential lectin inhibition of electrophysiological responses of the rat to some stimuli has also been reported,[39,40] the mechanism underlying these results was not studied. Thus, the inhibition of receptor binding in isolated cilia from catfish was consistent with the conclusion that lectin inhibition of electrophysiological responses probably resulted from inhibition at the level of the initial receptor binding event.

Alternative mechanisms, independent of specific stimulus-receptor interaction, have also been proposed to account for olfactory neuron activation.[25] In neuroblastoma cells, that presumably lack olfactory receptor proteins, membrane depolarization responses to several stimuli were obtained at concentrations that also altered membrane fluidity.[41] Similar odorant-dependent depolarizing responses were also observed in azolectin liposomes.[42] A correlation was noted between the minimum stimulus concentrations required for membrane depolarization and response thresholds in frog and porcine olfactory systems. Addition of sphingomyelin to azolectin or egg phosphatidylcholine liposomes increased the magnitude of nearly all stimulus-induced membrane responses, while cholesterol and phosphatidylethanolamine differentially affected fewer responses.[42,43] Thus, a model for olfactory reception was proposed in which membrane lipids provided adsorption sites for stimuli with no requirement for receptor protein. Membrane lipid composition was postulated to vary between individual receptor cells to account for the sensitivity and stimulus selectivity of olfactory responses. The postulated cellular lipid heterogeneity proposed in this model has not been confirmed experimentally. However, it is also not unreasonable to expect that some stimuli, particularly hydrophobic compounds, would alter membrane properties irrespective of the presence of protein components. Thus, in addition to stimulus interaction with receptor proteins, nonspecific adsorption onto, or partitioning of some stimuli into, the neuronal membrane may also contribute to the mechanism of olfactory neuron activation.

II. G PROTEINS IN OLFACTORY TRANSDUCTION

A. G-PROTEIN IDENTIFICATION AND LOCALIZATION

Recent progress in visual research, particularly with regard to the role of transducin in coupling photolyzed rhodopsin to cyclic GMP phosphodiesterase,[3] has encouraged speculation that chemosensory signal transduction may also be mediated by G proteins. Several laboratories have therefore recently tested the hypothesis that G-protein regulated second messenger systems are involved in olfactory receptor cell activation. G proteins were subsequently identified in the olfactory epithelium of several vertebrate species. The common β-subunit, M_r 35,000 to 36,000 was identified in isolated cilia preparations by immunoblotting analysis.[44,45] Immunohistochemical analysis with subunit specific antisera to the β-subunit and to a common amino acid sequence of G-protein α-subunits showed that G proteins were localized in the cilia and were also distributed throughout other regions of the receptor neuron membrane.[44] G protein α-subunits that have been identified in isolated olfactory cilia are listed in Table 1. In all species, a cholera toxin ADP-ribosylation substrate, M_r 45,000 to 42,000, was detected, corresponding to the α-subunit of G_s.[44,47] G_s is particularly abundant in olfactory cilia, is apparently absent from respiratory cilia,[44,47] and is immunochemically detectable in membranes derived from deciliated neuroepithelium.[44] Although cDNA clones have been identified from rat olfactory epithelium that encode both M_r 52,000 and 45,000 forms of the α-subunit of G_s,[48] only a single form of G_s has been detected in the olfactory epithelium of individual vertebrates by ADP-ribosylation or by immunoblotting analysis with subunit specific antisera.[44-47]

At least three distinct IAP ADP-ribosylation substrates were identified in isolated cilia preparations (Table 1). Immunoreactive transducin was not detected in the olfactory system.[44,45] In most species, G_i and G_o were identified based on electrophoretic mobility, sensitivity to IAP-catalyzed ADP-ribosylation, and/ or immunoreactivity.[18,44,46,47] Both G proteins were also detected in membranes derived from deciliated neuroepithelium,[44,46] and were also found in respiratory cilia by ADP-ribosylation,[47] but not by immunoblotting.[44] In contrast, in isolated cilia from catfish, a single IAP substrate was identified that migrated in polyacrylamide gels to a position intermediate between purified G_i and G_o α-subunits.[45] The IAP substrate did not cross-react with specific antisera to G_o, although G_o was readily identified in brain membranes from catfish. However, the M_r 40,000 IAP substrate did cross-react with antisera to a common amino acid sequence near the N-terminus of G-protein α-subunits. Since the properties of this G protein differed from those of other IAP substrates in olfactory cilia,[44,46,47] it was designated as G_{40} to distinguish it from G_i and G_o. Although the functional role of G_{40} in olfactory cilia has not been established, G protein α-subunits

TABLE 1
G-Protein α-Subunits Identified in Olfactory Cilia

α-Subunit	Mr($\times 10^{-3}$)[a]	ADP-Ribosylation[b]		Species	Ref.
		CT	IAP		
G$_s$	45,42	+	−	Frog	44,46
	42	+	−	Toad	47
	42	+	−	Rat	47
	45	+	−	Catfish	45
G$_i$	42,40	−	+	Frog	44,46
	40	−	+	Toad	47
	40	−	+	Rat	47
G$_o$	40,39	−	+	Frog	44,46
	39	−	+	Toad	47
	39	−	+	Rat	47
	40	−	+	Mouse	49
G$_{40}$	40	−	+	Catfish	45

[a]   M$_r$ by SDS-PAGE.
[b]   ADP-ribosylation catalyzed by cholera toxin (CT) or islet-activating protein (IAP).

with similar properties have also been reported in human neutrophils,[50] differentiated HL-60 cells,[51] and porcine brain.[52] It is not known whether G$_{40}$ is a unique G protein or a variant of G$_i$.[53,54] However, the possibility that G$_{40}$ corresponds to a G$_i$ variant (probably G$_i\alpha2$,[53,54]) seems likely, since cDNA clones have been identified in rat olfactory epithelium that encode three forms of G$_i$.[48]

B. G-PROTEIN REGULATION OF OLFACTORY TRANSDUCTION
1. G-Protein Coupling to Olfactory Receptors

The immunohistochemical localization of G proteins in the receptor cell membrane and their identification at the molecular level in the cilia suggested the likely participation of these proteins in olfactory signal transduction. The localization of G$_s$ in olfactory cilia, but not in respiratory cilia,[44,47] together with the observation of guanine nucleotide-dependent stimulation of adenylate cyclase by odorants in isolated cilia,[46] further implicated G-protein involvement in mediating transmembrane signaling in the cilia. It was therefore proposed that olfactory receptor occupation activated adenylate cyclase by the classical G$_s$-mediated mechanism.[2,46,53] That olfactory receptors were actually coupled to G proteins in the assumed manner was experimentally confirmed by ligand binding studies in isolated cilia from catfish.[45] The affinities of two olfactory L-amino acid receptors for their ligands were decreased by about an order of magnitude in the presence of GTP or the hydrolysis-resistant analogue Gpp(NH)p, indicating that these receptors were coupled to G proteins in the classical manner described for hormone and neurotransmitter receptors. Additional evidence, described below, derived from second messenger studies, has also implicated G-protein involvement in olfactory transduction. However, the shift in receptor affinity induced by guanine nucleotides documented for the first time, functional interactions between the receptors and G proteins at the level of the initial binding event.

2. Adenylate Cyclase

A role for adenylate cyclase in olfactory transduction was initially suggested by the observation of unusually high basal levels of the enzyme in the olfactory epithelium.[55] Membrane permeable cyclic AMP analogs, but not the corresponding cyclic GMP analogs, and cyclic nucleotide phosphodiesterase inhibitors reversibly reduced the amplitude of stimulus-evoked electrophysiological responses.[56] Since these initial experiments, biochemical characterization and stimulus activation of adenylate cyclase were investigated in isolated cilia and membrane preparations from frog,[46,47,57] toad,[47] rat,[47,57,58] and catfish.[59] The olfactory adenylate cyclase exhibited several properties of the classical hormone-regulated enzyme,[53] including sensitivity to forskolin and cholera toxin.[46,47,57,59] Cyclic AMP formation was also increased by established G-protein effectors, including fluoride ion, GTP and its hydrolysis-resistant analogs, and GDP, while GDPβS inhibited basal adenylate cyclase activity. Stimulus activation of adenylate cyclase in isolated cilia was GTP-dependent, further implicating G-protein participation, presumably G_s, in mediating activation of the olfactory enzyme. A role for G_s in human olfaction was also suggested by the observation of impaired olfactory ability in type 1a pseudohypoparathyroidism patients.[60] These patients exhibited decreased G_s activity in erythrocytes and resistance to the cyclic AMP-mediated actions of several hormones. In contrast, type 1b patients had normal G_s activity and olfactory ability.

The potential second messenger function of cyclic AMP in olfactory neurons was also investigated biochemically and by electrophysiological techniques. Cyclic nucleotide-dependent protein kinase activity was identified in isolated cilia from frog using exogenous histone substrate.[61] The enzyme was activated by both cyclic AMP and cyclic GMP, although half-maximal stimulation required about tenfold more cyclic GMP than cyclic AMP. Endogenous protein substrates for the kinase were also identified in isolated cilia from frog[61] and bovine[62] olfactory epithelium. In both species, ciliary-specific phosphoprotein, pp24, exhibited specifically enhanced phosphorylation in response to cyclic AMP. The identity of this phosphoprotein and its possible role in signal transduction were not determined. Electrophysiological evidence also indicated that cyclic nucleotides modulate ion channel activity in olfactory neurons. Olfactory epithelium homogenates incorporated into planar phospholipid bilayers conferred odorant sensitivity to the bilayers that was ATP- and GTP-dependent and mimicked by cyclic AMP.[63] In these preparations, exogenous ATP decreased channel open time. Since protein kinase inhibitor antagonized the effect of ATP, it was concluded that cyclic AMP-regulated channels were modulated directly by the nucleotide as well as by protein phosphorylation.[64] Cyclic nucleotide modulation of membrane conductance in excised membrane patches from individual cilia on dissociated olfactory neurons was also reported.[12] In this study, it was concluded that the cation channels mediating the conductance were directly gated by cyclic nucleotides without protein phosphorylation since exogenous nucleotide triphosphates were not required to observe the reversible cyclic nucleotide-dependent conductance change. Similar results were also obtained in isolated cilia incorporated into phospholipid bilayers on the tips of patch-clamp electrodes.[59] Cyclic AMP and cyclic GMP were equally effective in modulating cation channel activity in both excised ciliary patches[12] and in cilia reconstituted in artificial membranes.[59]

Taken together, biochemical and neurophysiological evidence is consistent with the hypothesis that adenylate cyclase is involved in olfactory signal transduction. Stimulus interaction with receptor would subsequently activate the enzyme, presumably by the classical G_s-dependent mechanism.[2] The resulting increased cyclic AMP levels, by direct modulation of ion channel activity[12,59] and/or by stimulation of cyclic nucleotide-dependent protein kinase,[61,64] would elicit the observed increase in membrane conductance. While a cyclic AMP-mediated cascade may account for responses to some odorants, it is likely that additional mechanisms are involved in olfactory stimulus-response coupling since some odorants do not stimulate adenylate cyclase.[57] Since many odorants that stimulate cyclic AMP formation are hydrophobic[57] and alter membrane fluidity at similar concentrations required to activate adenylate cyclase,[41-43] the possibility that nonspecific membrane perturbation may contribute to the observed increases in cyclic AMP levels has not been critically evaluated. A recent study indicated that the olfactory adenylate cyclase of rat was relatively insensitive to temperature change and to benzyl alcohol, an agent frequently used to perturb membrane fluidity.[65] Since neither ligand binding nor membrane fluidity measurements were reported, the interpretation of these observations is limited.

Amino acid stimulation of cyclic AMP formation was also studied in isolated cilia from catfish under conditions that maintained receptor binding activity and receptor coupling to G proteins.[59] Significant increases in cyclic AMP levels were observed in the presence of guanine nucleotide only at high receptor occupancy levels after prolonged (greater than 2 min) exposure to stimulus. In addition, for ten amino acids, little correlation was obtained between receptor specificity or electrophysiological potency and the ability to stimulate adenylate cyclase. These observations suggested that cyclic AMP may be involved in tonic or adaptive responses resulting from high stimulus concentration or prolonged exposure to stimulus, rather than mediating the initial, much faster (less than 1 s), phasic response. A similar conclusion was reached by Pace et al.[66] based on the observation of decreased stimulus-dependent cyclic AMP formation following prior exposure to stimulus.

3. Phosphoinositide Hydrolysis

Although less extensively investigated than adenylate cyclase, the possibility that phosphoinositide-derived second messengers may be involved in olfaction was also studied recently. Phospholipase C (phosphatidylinositol-4,5-bisphosphate phosphodiesterase) activity was detected in the olfactory epithelium of catfish, mouse and rat.[67,68] As in many other tissues, the majority of the enzyme activity was recovered in the soluble fraction, although about 5% of the total activity was reproducibly associated with the isolated cilia.[20] Not surprisingly, the olfactory enzyme appears to be heterogeneous as indicated by the resolution of multiple molecular species by gel filtration and the observation of two pH optima. Stimulus activation of the enzyme was initially demonstrated in isolated cilia from catfish.[67,68] Amino acids rapidly stimulated two- to three-fold increases in inositol phosphate formation from exogenous PIP_2 within 10 s of exposure to stimulus. The likely participation of a G protein in mediating activation of phospholipase C was suggested by stimulation of inositol phosphate production by GTP and its hydrolysis-resistant analogs. Stimulus activation of phosphoinositide hydrolysis was also GTP-dependent, further implicating G-protein involvement in coupling of the olfactory amino acid receptors to phospholipase C. The identity of the G protein mediating activation of phosphoinositide hydrolysis in the cilia has not been determined.

Anion-exchange chromatography of inositol phosphates formed under basal conditions or following incubation with guanine nucleotides and amino acids indicated that IP_3 was the major product.[20,68] Although the isomeric composition of the ''IP_3'' fraction was not determined, the major product was presumably the 1,4,5-isomer. The potential second messenger function of IP_3 in mediating Ca^{2+} release from intracellular stores was also investigated in isolated microsomes from the olfactory epithelium. The isolated microsomal preparations sequestered Ca^{2+} in a Mg^{2+}/ATP-dependent manner (Figure 1A). Part of the ATP-dependent Ca^{2+} uptake was inhibited by oligomycin, an inhibitor of mitochondrial Ca^{2+} uptake. About 90% of the sequestered Ca^{2+} was released by the calcium ionophore A23187 (Figure 1B). IP_3 rapidly released about 30% of the accumulated Ca^{2+}. IP_3-mediated Ca^{2+} release was transient and was followed by subsequent re-uptake (Figure 1B). Although the functional consequences of IP_3-mediated Ca^{2+} release in isolated microsomes from the olfactory epithelium have not been determined, it is of interest to note that the olfactory adenylate cyclase is inhibited by Ca^{2+}.[57,58] Thus, sufficient Ca^{2+} may be released from internal stores by IP_3 to provide an additional mechanism of regulation of cyclic AMP production. In addition, Ca^{2+} release, in conjunction with diacylglycerol production, may serve to activate protein kinase C. Protein kinase C has been identified in isolated cilia from frog by immunoreactivity and phorbol ester binding,[44] although activation of the enzyme with Ca^{2+} and phospholipid was not achieved in another study.[61]

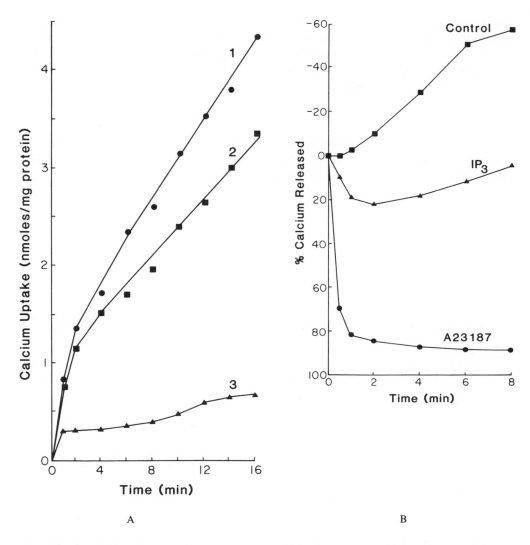

FIGURE 1.   Time course of Ca^{2+} uptake and release in isolated microsomes from the olfactory epithelium. (A) Ca^{2+} accumulation into microsomes was determined with $^{45}Ca^{2+}$ in the presence (curves 1 and 2) or absence (curve 3) of 1 mM ATP and 5 mM $MgCl_2$.[69,70] External Ca^{2+} was buffered with EGTA to 1 μM. Curve 2 shows the ATP-dependent Ca^{2+} uptake in the presence of 10 μM oligomycin. (B) Ca^{2+} release from microsomes obtained with 5 μM IP_3 and 5 μM A23187. The data are expressed as the percent of accumulated Ca^{2+} released in the presence of each effector.

III. CONCLUSIONS

Sufficient evidence has accumulated recently to support the hypothesis of second messenger involvement in olfactory signal transduction. The participation of G proteins in mediating regulation of second messenger events in olfactory neurons following stimulus-receptor interaction has also been indicated, since stimulus activation of both cyclic AMP and inositol phosphate formation are GTP-dependent. While the functional role of G_s in mediating stimulation of adenylate cyclase is assumed, the role of other identified G proteins in olfactory transduction is less clear. In particular, it is tempting to speculate that one or more IAP substrates in olfactory neurons may be involved in direct regulation of ion channel activity as recently described in heart.[71] Another attractive hypothesis is the possibility that the unusual G_{40} IAP substrate may be involved in coupling olfactory receptors to phosphoinositide hydrolysis as recently suggested for the chemotactic peptide receptor in leukocytes.[72] Confirmation of these hypotheses, as well as identification of additional functions of G proteins in regulation of receptor activation and second messenger events, will substantially advance our understanding of transmembrane signaling in olfactory neurons.

ACKNOWLEDGMENTS

Work in the author's laboratory was supported by the National Science Foundation, National Institutes of Health, and the Veterans Administration. The author thanks Mrs. Janice Blescia for efficient and expert preparation of the manuscript.

REFERENCES

1. **Getchell, T. V.**, Functional properties of vertebrate olfactory receptor neurons, *Physiol. Rev.*, 66, 772, 1986.
2. **Lancet, D.**, Vertebrate olfactory reception, *Annu. Rev. Neurosci.*, 9, 329, 1986.
3. **Stryer, L.**, Cyclic GMP cascade of vision, *Annu. Rev. Neurosci.*, 9, 87, 1986.
4. **Cagan, R. H. and Kare, M. R.**, Eds., *Biochemistry of Taste and Olfaction*, Academic Press, New York, 1981.
5. **Getchell, T. V., Margolis, F. L., and Getchell, M. L.**, Perireceptor and receptor events in vertebrate olfaction, *Prog. Neurobiol.*, 23, 317, 1985.
6. **Yamamoto, M.**, Comparative morphology of the peripheral olfactory organs in teleosts, in *Chemoreception in Fishes*, Hara, T. J., Ed., Elsevier, New York, 1982, 39.
7. **Moran, D. T., Rowley, J. C., and Jafek, B. W.**, Electron microscopy of human olfactory epithelium reveals a new cell type: the microvillar cell, *Brain Res.*, 253, 39, 1982.
8. **Thommesen, G.**, Morphology, distribution and specificity of olfactory receptor cells in salmonid fishes, *Acta Physiol. Scand.*, 117, 241, 1983.
9. **Erickson, J. R. and Caprio, J.**, The spatial distribution of ciliated and microvillous olfactory receptor neurons in the channel catfish is not matched by a differential specificity to amino acid and bile salt stimuli, *Chem. Senses*, 9, 127, 1984.
10. **Bruch, R. C., Kalinoski, D. L., and Kare, M. R.**, Biochemistry of vertebrate olfaction and taste, *Annu. Rev. Nutr.*, 8, 21, 1988.
11. **Maue, R. A. and Dionne, V. E.**, Preparation of isolated mouse olfactory receptor neurons, *Pflugers Arch.*, 409, 244, 1987.
12. **Nakamura, T. and Gold, G. H.**, A cyclic nucleotide-gated conductance in olfactory receptor cilia, *Nature*, 325, 442, 1987.
13. **Trotier, D.**, A patch-clamp analysis of membrane currents in salamander olfactory receptor cells, *Pflugers Arch.*, 407, 589, 1986.
14. **Hirsch, J. D. and Margolis, F. L.**, Isolation, separation, and analysis of cells from olfactory epithelium, in *Biochemistry of Taste and Olfaction*, Cagan, R. H. and Kare, M. R., Eds., Academic Press, New York, 1981, 311.
15. **Kleene, S. J. and Gesteland, R. C.**, Dissociation of frog olfactory epithelium with *N*-ethylmaleimide, *Brain Res.*, 229, 536, 1981.
16. **Rhein, L. D. and Cagan, R. H.**, Role of cilia in olfactory recognition, in *Biochemistry of Taste and Olfaction*, Cagan, R. H. and Kare, M. R., Eds., Academic Press, New York, 1981, 47.
17. **Rhein, L. D. and Cagan, R. H.**, Biochemical studies of olfaction: isolation, characterization, and odorant binding activity of cilia from rainbow trout olfactory rosettes, *Proc. Natl. Acad. Sci. U.S.A.*, 77, 4412, 1980.
18. **Anholt, R. R. H., Aebi, U., and Snyder, S. H.**, A partially purified preparation of isolated chemosensory cilia from the olfactory epithelium of the bullfrog, *Rana catesbeiana, J. Neurosci.*, 6, 1962, 1986.

19. **Chen, Z., Pace, U., Heldman, J., Shapira, A., and Lancet, D.,** Isolated frog olfactory cilia: a preparation of dendritic membranes from chemosensory neurons, *J. Neurosci.,* 6, 2146, 1986.
20. **Boyle, A. G., Park, Y. S., Huque, T., and Bruch, R. C.,** Properties of phospholipase C in isolated cilia from the channel catfish *(Ictalurus punctatus), Comp. Biochem. Physiol.,* 88B, 767, 1987.
21. **Chen, Z. and Lancet, D.,** Membrane proteins unique to vertebrate olfactory cilia: candidates for sensory receptor molecules, *Proc. Natl. Acad. Sci. U.S.A.,* 81, 1859, 1984.
22. **Chen, Z., Ophir, D., and Lancet, D.,** Monoclonal antibodies to ciliary glycoproteins of frog olfactory neurons, *Brain Res.,* 368, 329, 1986.
23. **Chen, Z., Pace, U., Ronen, D., and Lancet, D.,** Polypeptide gp95: a unique glycoprotein of olfactory cilia with transmembrane receptor properties, *J. Biol. Chem.,* 261, 1299, 1986.
24. **Price, S.,** Receptor proteins in vertebrate olfaction, in *Biochemistry of Taste and Olfaction,* Cagan, R. H. and Kare, M. R., Eds., Academic Press, New York, 1981, 69.
25. **Kurihara, K., Yoshii, K., and Kashiwayanagi, M.,** Transduction mechanisms in chemoreception, *Comp. Biochem. Physiol.,* 85A, 1, 1986.
26. **Price, S. and Willey, A.,** Benzaldehyde binding protein from dog olfactory epithelium, *Chem. Senses,* 11, 651, 1986.
27. **Pelosi, P., Baldaccini, E., and Pisanelli, A. M.,** Identification of a specific olfactory receptor for 2-isobutyl-3-methoxypyrazine, *Biochem. J.,* 201, 245, 1982.
28. **Pevsner, J., Trifiletti, R. R., Strittmatter, S. M., and Snyder, S. H.,** Isolation and characterization of an olfactory receptor protein for odorant pyrazines, *Proc. Natl. Acad. Sci. U.S.A.,* 82, 3050, 1985.
29. **Pevsner, J., Sklar, P. B., and Snyder, S. H.,** Localization of odorant binding protein (OBP) to nasal glands and secretions, *Chem. Senses,* 11, 650, 1986.
30. **Lee, K. H., Wells, R. G., and Reed, R. R.,** Isolation of an olfactory cDNA: similarity to retinol-binding protein suggests a role in olfaction, *Science,* 235, 1053, 1987.
31. **Cagan, R. H. and Zeiger, W. N.,** Biochemical studies of olfaction: binding specificity of radioactively labeled stimuli to an isolated olfactory preparation from rainbow trout *(Salmo gairdneri), Proc. Natl. Acad. Sci. U.S.A.,* 75, 4679, 1978.
32. **Rhein, L. D. and Cagan, R. H.,** Biochemical studies of olfaction: binding specificity of odorants to a cilia preparation from rainbow trout olfactory rosettes, *J. Neurochem.,* 41, 569, 1983.
33. **Brown, S. B. and Hara, T. J.,** Biochemical aspects of amino acid receptors in olfaction and taste, in *Chemoreception in Fishes,* Hara, T. J., Ed., Elsevier, New York, 1982, 159.
34. **Rehnberg, B. G. and Schreck, C. B.,** The olfactory L-serine receptor in coho salmon: biochemical specificity and behavioral response, *J. Comp. Physiol.,* 159, 61, 1986.
35. **Kalinoski, D. L., Bruch, R. C., and Brand, J. G.,** Differential interaction of lectins with chemosensory receptors, *Brain Res.,* 418, 34, 1987.
36. **Bruch, R. C. and Rulli, R. D.,** Ligand binding specificity of a neutral L-amino acid olfactory receptor, *Comp. Biochem. Physiol.,* 91B, 535, 1988.
37. **Caprio, J.,** Olfaction and taste in the channel catfish: an electrophysiological study of the responses to amino acids and derivatives, *J. Comp. Physiol.,* 123, 357, 1978.
38. **Caprio, J. and Byrd, R. P., Jr.,** Electrophysiological evidence for acidic, basic, and neutral amino acid olfactory receptor sites in the catfish, *J. Gen. Physiol.,* 84, 403, 1984.
39. **Shirley, S. G., Polak, E. H., Mather, R. A., and Dodd, G. H.,** The effect of concanavalin A on the rat electro-olfactogram: differential inhibition of odorant response, *Biochem. J.,* 245, 175, 1987.
40. **Shirley, S. G., Polak, E. H., Edwards, D. A., Wood, M. A., and Dodd, G. H.,** The effect of concanavalin A on the rat electro-olfactogram at varying odorant concentrations, *Biochem. J.,* 245, 185, 1987.
41. **Kashiwayanagi, M. and Kurihara, K.,** Evidence for non-receptor odor discrimination using neuroblastoma cells as a model for olfactory cells, *Brain Res.,* 359, 97, 1985.
42. **Nomura, T. and Kurihara, K.,** Liposomes as a model for olfactory cells: changes in membrane potential in response to various odorants, *Biochemistry,* 26, 6135, 1987.
43. **Nomura, T. and Kurihara, K.,** Effects of changed lipid composition on response of liposomes to various odorants: possible mechanism of odor discrimination, *Biochemistry,* 26, 6141, 1987.
44. **Anholt, R. R. H., Mumby, S. M., Stoffers, D. A., Girard, P. R., Kuo, J. F., and Snyder, S. H.,** Transduction proteins of olfactory receptor cells: identification of guanine nucleotide binding proteins and protein kinase C, *Biochemistry,* 26, 788, 1987.
45. **Bruch, R. C. and Kalinoski, D. L.,** Interaction of GTP-binding regulatory proteins with chemosensory receptors, *J. Biol. Chem.,* 262, 2401, 1987.
46. **Pace, U., Hanski, E., Salomon, Y., and Lancet, D.,** Odorant-sensitive adenylate cyclase may mediate olfactory reception, *Nature,* 316, 255, 1985.
47. **Pace, U. and Lancet, D.,** Olfactory GTP-binding protein: signal-transducing polypeptide of vertebrate chemosensory neurons, *Proc. Natl. Acad. Sci. U.S.A.,* 83, 4947, 1986.
48. **Jones, D. T. and Reed, R. R.,** Molecular cloning of five GTP-binding protein cDNA species from rat olfactory neuroepithelium, *J. Biol. Chem.,* 262, 14241, 1987.

49. **Bruch, R. C.**, unpublished data, 1987.
50. **Gierschik, P., Falloon, J., Milligan, G., Pines, M., Gallin, J. I., and Spiegel, A.**, Immunochemical evidence for a novel pertussis toxin substrate in human neutrophils, *J. Biol. Chem.*, 261, 8058, 1986.
51. **Oinuma, M., Katada, T., and Ui, M.**, A new GTP-binding protein in differentiated human leukemic (HL-60) cells serving as the specific substrate of islet-activating protein, pertussis toxin, *J. Biol. Chem.*, 262, 8347, 1987.
52. **Katada, T., Oinuma, M., Kusakabe, K., and Ui, M.**, A new GTP-binding protein in brain tissues serving as the specific substrate of islet-activating protein, pertussis toxin, *FEBS Lett.*, 213, 353, 1987.
53. **Gilman, A. G.**, G-proteins: transducers of receptor-generated signals, *Annu. Rev. Biochem.*, 56, 615, 1987.
54. **Goldsmith, P., Gierschik, P., Milligan, G., Unson, C., Vinitsky, R., Malech, H. L., and Spiegel, A. M.**, Antibodies directed against synthetic peptides distinguish between GTP-binding proteins in neutrophil and brain, *J. Biol. Chem.*, 262, 14683, 1987.
55. **Kurihara, K. and Koyama, N.**, High activity of adenyl cyclase in olfactory and gustatory organs, *Biochem. Biophys. Res. Commun.*, 48, 30, 1972.
56. **Menevse, A., Dodd, G., and Poynder, T. M.**, Evidence for the specific involvement of cyclic AMP in the olfactory transduction mechanism, *Biochem. Biophys. Res. Commun.*, 77, 671, 1977.,
57. **Sklar, P. B., Anholt, R. R. H., and Snyder, S. H.**, The odorant-sensitive adenylate cyclase of olfactory receptor cells: differential stimulation by distinct classes of odorants, *J. Biol. Chem.*, 261, 15538, 1986.
58. **Shirley, S. G., Robinson, C. J., Dickinson, K., Aujla, R., and Dodd, G. H.**, Olfactory adenylate cyclase of the rat: stimulation by odorants and inhibition by Ca^{2+}, *Biochem. J.*, 240, 605, 1986.
59. **Bruch, R. C. and Teeter, J. H.**, Role of cyclic nucleotides in olfactory signal transduction, *J. Biol. Chem.*, submitted.,
60. **Weinstock, R. S., Wright, H. N., Spiegel, A. M., Levine, M. A., and Moses, A. M.**, Olfactory dysfunction in humans with deficient guanine nucleotide-binding protein, *Nature*, 322, 635, 1986.
61. **Heldman, J. and Lancet, D.**, Cyclic AMP-dependent protein phosphorylation in chemosensory neurons: identification of cyclic nucleotide-regulated phosphoproteins in olfactory cilia, *J. Neurochem.*, 47, 1527, 1986.
62. **Kropf, R., Lancet, D., and Lazard, D.**, A bovine olfactory cilia preparation: specific transmembrane glycoproteins and phosphoproteins, *Soc. Neurosci, Abstr.*, 13(2), 1410, 1987.
63. **Vodyanoy, V. and Vodyanoy, I.**, ATP and GTP are essential for olfactory response, *Neurosci. Lett.*, 73, 253, 1987.
64. **Vodyanoy, V. and Vodyanoy, I.**, Ion channel modulation by cAMP and protein kinase inhibitor, *Soc. Neurosci. Abstr.*, 13(2), 1410, 1987.
65. **Shirley, S. G., Robinson, C. J., and Dodd, G. H.**, The influence of temperature and membrane fluidity changes on the olfactory adenylate cyclase of the rat, *Biochem. J.*, 245, 613, 1986.
66. **Pace, U., Heldman, J., Shafir, I., Rimon, G., and Lancet, D.**, Molecular correlates of olfactory adaptation: adenylate cyclase and protein phosphorylation, *Soc. Neurosci. Abstr.*, 13(1), 362, 1987.
67. **Huque, T. and Bruch, R. C.**, Odorant- and guanine nucleotide-stimulated phosphoinositide turnover in olfactory cilia, *Biochem. Biophys. Res. Commun.*, 137, 36, 1986.
68. **Bruch, R.C., Kalinoski, D. L., and Huque, T.**, Role of GTP-binding regulatory proteins in receptor-mediated phosphoinositide turnover in olfactory cilia, *Chem. Senses*, 12, 173, 1987.
69. **Ueda, T., Chueh, S. H., Noel, M. W., and Gill, D. L.**, Influence of inositol 1,4,5-trisphosphate and guanine nucleotides on intracellular calcium release within the NIE-115 neuronal cell line, *J. Biol. Chem.*, 261, 3184, 1986.
70. **Jean, T and Klee, C. B.**, Calcium modulation of inositol 1,4,5-trisphosphate-induced calcium release from neuroblastoma × glioma hybrid (NG108-15) microsomes, *J. Biol. Chem.*, 261, 16414, 1986.
71. **Yatani, A., Codina, J., Brown, A. M., and Birnbaumer, L.**, Direct activation of mammalian atrial muscarinic potassium channels by GTP regulatory protein G_k, *Science*, 235, 207, 1987.
72. **Uhing, R. J., Polakis, P. G., and Snyderman, R.**, Isolation of GTP-binding proteins from myeloid HL-60 cells: identification of two pertussis toxin substrates, *J. Biol. Chem.*, 262, 15575, 1987.

Chapter 8

THE RELEASE OF CALCIUM BY LIGHT IN THE PHOTORECEPTORS OF INVERTEBRATES

Richard Payne and Alan Fein

TABLE OF CONTENTS

I.   Introduction 136
        A.   Rhodopsin, the Receptor for Light 136
        B.   Rhodopsin's Special Environment: the Microvillus 137

II.   Rhodopsin is a Member of a Family of Receptors Thought to Activate
        G-Proteins 138
        A.   Evidence for Light-Activated G-Proteins in Microvillar
                Photoreceptors 139
                1.   Light-Activated GTP-ase Activity 140
                2.   Light-Activated Binding of GTP-Analogs to Microvillar
                        Membrane 140
        B.   The Identification of G Proteins in Microvillar Membranes 140
                1.   Toxin-Catalyzed Labels 140
                2.   GTP Photoaffinity Labels 141

III.   Physiology of Microvillar Photoreceptors 142
        A.   Physiological Evidence for the Involvement of G Proteins
                in Phototransduction 142
        B.   Release of Calcium from Internal Stores by Light 143
        C.   The Role of Calcium Ions in the Response to Light 144
        D.   Inositol 1,4,5-triphosphate Mediates Calcium Release 145
        E.   Evidence for the Light-Induced Production of Ins(1,4,5)P$_3$ 147
        F.   The Control of Phospholipase C by Rhodopsin 148

IV.   Summary 149

References 150

I. INTRODUCTION

The photoreceptors of invertebrates are structurally and physiologically different from those of vertebrates. A prominent characteristic of the phototransduction process in the photoreceptors of invertebrates is that the electrical response of the photoreceptor is accompanied by a rapid rise in the intracellular concentration of free calcium ions (Ca_i).[1] Retinal rod photoreceptors of vertebrates do not display a similar rise in Ca_i. On the contrary, there is now convincing evidence that Ca_i in rods falls on illumination.[2-4] However, although the changes in Ca_i may differ, the cascade of biochemical reactions mediating phototransduction in photoreceptors of both vertebrates and invertebrates may share an important similarity. In both cases, phototransduction appears to be initiated via the activation of a G protein by the photoactivated visual pigment, rhodopsin. As we shall show, the photoreceptors of invertebrates are an excellent system in which to investigate the linkage of the mobilization of intracellular calcium to the activation of a G protein. The eyes of invertebrates contain high concentrations of receptor molecules (rhodopsin) and G proteins in specialized organelles, enabling biochemical analysis of phototransduction, while the large size of some of the photoreceptors enables physiological and pharmacological experiments on intact cells as well. In this review, we examine the proposal that the photoreceptors of invertebrates are specialized examples of a general mechanism by which receptors mobilize calcium via the production of inositol triphosphate.

A. RHODOPSIN, THE RECEPTOR FOR LIGHT

The general name, rhodopsin, will be used in this review for all of the visual pigments of the photoreceptors of invertebrates. Rhodopsin consists of a chromophore (11-*cis* retinal or an analog) that is covalently bound to the apoprotein opsin (see review by Kirchfeld[5]). Light initiates the visual process by the photoisomerisation of the chromophore from an 11-*cis* to all-*trans* configuration. This transformation of chromophore structure alters the conformation of the surrounding opsin moiety.

Opsin (MW 37 to 46 kDa) spans the plasma membrane of the photoreceptor (Figure 1) (reviewed by Findlay[6] and Applebury and Hardgrave[7]). The interaction between the chromophore and opsin tunes the maximal absorption of the chromophore to a particular wavelength of light within a spectral range of 350 to 540 nm (reviewed by Hamdorf[8]). Several species of invertebrates have acute color vision and their eyes contain photoreceptors having differing spectral sensitivities. For example, there are three anatomical classes among the eight photoreceptors that lie under every facet in the compound eye of the fly *Drosophila*. Each class has a peak spectral sensitivity at a different wavelength;[9] six of the receptors (R1 to 6) are maximally sensitive to blue-green light, one (R7) to ultraviolet light, and one (R8) to blue light. Since opsin structure determines, in part, the wavelength of maximal absorption of the visual pigment, this diversity of spectral sensitivities indicates a diversity of opsin structure within a given organism. Recently, four different *Drosophila* genes have been sequenced, each of which presumably codes for an opsin. One gene is exclusively expressed in R1 to 6.[10,11] Surprisingly, there are two genes expressed in nonoverlapping populations of R7 receptors,[13,14] probably implying a spectral diversity within the anatomical class of R7 receptors of *Drosophila* similar to that known to exist in other flies.[15]

Comparison of the structures of the opsins expressed in *Drosophila* enables an identification of conserved regions that are presumably important for function (Figure 1). All sequences show seven hydrophobic regions, which, by analogy with vertebrate opsin, span the membrane as alpha helices.[7,16] The chromophore in *Drosophila*, 11-*cis* 3-hydroxyretinal,[17,18] is linked via a Schiff base onto a lysine residue situated in the middle of transmembrane helix VII. This lysine residue is conserved in all opsins so far studied, both from vertebrates and invertebrates.[7,14]

Cytoplasmic

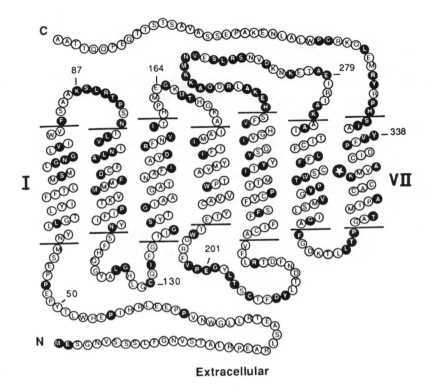

Extracellular

FIGURE 1. The proposed structure of one of the opsin genes (Rh3) of the fly *Drosophila melanogaster*. Amino acid residues are indicated by their single letter codes. Black solid circles indicate identities among the four Drosophila opsin genes so far sequenced. The roman numerals indicate the numbering of the transmembrane α-helices. The cytoplasmic loops between helices I to II and V to VI show extensive sequence homologies within the *Drosophila* family of opsins. The residue on helix VII that probably binds the chromophore is indicated by a star. (From Zucker, C. S., Montell, C., Jones, K., Laverty, T., and Rubin, G. M., *J. Neurosci.*, 7, 1550, 1987. With permission.)

When photoisomerization of the chromophore occurs, a stable product is formed that is spectroscopically distinct from rhodopsin and is called metarhodopsin. Metarhodopsin is formed within 1 ms of the absorption of a photon by rhodopsin.[19] A conformational change in the opsin moiety somehow caused by the alteration in chromophore structure is presumably transmitted from the hydrophobic core of opsin to the cytoplasmic surface, enabling interaction of metarodopsin with a G protein (see below) and a kinase that phosphorylates metarhodopsin.[20,21] So far, these are the only known biochemical interactions of metarhodopsin. Two loops (I-II) and (V-VI) on the cytoplasmic side are highly conserved amongst the opsins of *Drosophila*, suggesting some functional significance for the interaction of metarhodopsin with GTP-binding proteins (see below). The C terminus, situated on the cytoplasmic side, contains serine and threonine residues which are potential sites for phosphorylation.[22]

B. RHOSOPSIN'S SPECIAL ENVIRONMENT: THE MICROVILLUS

In the photoreceptors of many invertebrates, rhodopsin is concentrated in the membrane of structures called microvilli. Microvilli are cylindrical foldings of the plasmalemma, 50 to 80 nm in diameter and typically 1 to 2 μm in length. The microvilli are often closely packed along one surface of a photoreceptor to form a light-absorbing rhabdom, through which light is funneled from the lens of the eye. A typical photoreceptor bears about 10^5 microvilli, which constitute a large surface area for absorbing light, making the photoreceptor very sensitive. The long axes of the microvilli of a photoreceptor are typically aligned with each other, resulting in the preferential absorption by rhodopsin of light polarized along the axis of alignment.[23,24] Like the rhodopsin-bearing discs of the rod photoreceptors of vertebrates, microvilli contain a high percentage of protein. The ratio of protein:lipid in cephalopod microvillar membrane is approximately 50:50 wt%. Of the total protein, 50 to 70% is rhodopsin.[25-28] Rhodopsin is densely packed in the microvillus, since each microvillus contains many thousand molecules. In squid microvilli there is only a 5 to 6 nm separation between neighboring rhodopsin molecules.[28] The other proteins associated with the microvillar membrane are those thought to be intermediates in the transduction process (see below) and an extensive cytoskeletal network that supports the microvilli. The cytoskeleton appears to consist of two structural forms. Internally, a 6 to 12 nm diameter axial filamentous core runs the length of the microvillus, made up, probably, of one or several actin filaments.[29-32] Side arms may link this filament to the microvillar membrane.[30,31] In addition to this internal cytoskeleton, the outer surfaces of neighbouring microvilli appear to be joined together by proteins at discrete points along the microvillus length.[28] The microvilli, therefore, form a tight, cohesive bundle and the rhodopsin molecules within the cylindrical lipid bilayer of the microvillus are interspersed with cytoskeletal proteins on both the extracellular and the intracellular surfaces of the membrane.

As regards the lipid environment of rhodopsin, polyunsaturated phospholipids account for 62 to 85 mol% of the lipid and the majority of the remaining lipid is cholesterol (10 to 29 mol%).[25-27] The high degree of unsaturation of the phospholipids makes it likely that the lipids in the membrane are in a fluid state. Electron spin resonance spectra of membrane probes and direct measurement of the diffusion of fluorescent lipid analogs confirm the fluidity of the lipids, although neither method could determine whether the microvillar membrane is uniformly fluid.[27,35] Despite the apparent fluidity of the lipids, no translational diffusion of rhodopsin has been found within crayfish microvilli or between squid microvilli ($D \leqslant 10^{-9}$ cm^2/s).[37,38] The reason for the restricted mobility of rhodopsin is unknown, but it seems possible that the rhodopsin is anchored or constrained in some way by attachments to the cytoskeleton.

II. RHODOPSIN IS A MEMBER OF A FAMILY OF RECEPTORS THOUGHT TO ACTIVATE G PROTEINS

Sequence homologies between rhodopsins and other receptor molecules suggest that *Drosophila* opsin is a member of a family of receptor proteins, each of which initiates a different enzyme cascade via the activation of specific G proteins. Other members include the muscarinic acetylcholine receptor cloned by Kubo et al.,[39] the β adrenergic receptor,[40] and vertebrate opsin[122,123] (see review by Hall[41]). Photoactivation of the rhodopsin in the rod photoreceptors of vertebrates initiates a reduction in levels of cGMP in the rod, causing cGMP-activated channels to close.[42] The β adrenergic receptor initiates the activation of an adenylate cyclase to increase intracellular levels of cAMP in a variety of tissues (see review by Levitzki[43]). The muscarinic acetylcholine receptor cloned by Kubo et al.[34] probably activates a phospholipase C so as to increase intracellular levels of inositol 1,4,5 triphosphate ($InsP_3$).

The similarity of the receptors in this family is reinforced when one is substituted for another under artificial conditions. Photoactivated rhodopsins from invertebrates can cross-react with transducin, the G protein that is normally activated by rod rhodopsin and so initiate the hydrolysis of cGMP.[44-46] More impressively, photoactivated rod rhodopsin can

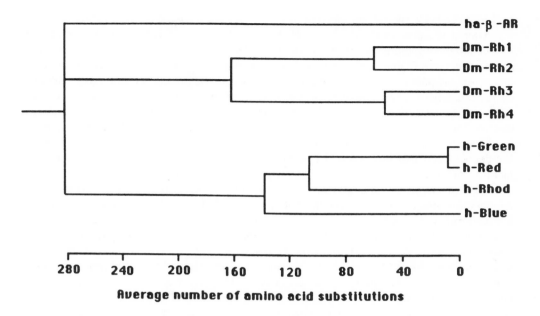

FIGURE 2. Phylogeny of visual pigments and the β-adrenergic receptor, constructed on the basis of the principle of minimal mutation distances. Number of mutational events are indicated in the scale beneath the tree. Dm-Rh1 stands for the Drosophila ninaE gene (the pigment of R1-6), Dm-Rh2, Dm-Rh3, and Dm-Rh4 refer to the pigments of the ocelli and the two of R7, respectively. The hamster β adrenergic receptor gene[40] is referred to as ha-β-AR. h-Red, h-Blue, h-Green refer to the human cone opsins and h-Rhod to human rhodopsin.[122,123] (From Zucker, C. S., Montell, C., Jones, K., Laverty, T., and Rubin, G. M., *J. Neurosci.*, 7, 1550, 1987. With permission.)

activate the G protein that confers hormone-sensitivity to adenylate cyclase.[47] The receptors of this family share the seven helical transmembrane segments of rhodopsin and a striking conservation of the amino acids in loop I-II on the cytoplasmic side, suggesting possible significance of this loop for G-protein activation (see review by Hall[41]). However, there are not enough examples of receptors known to be associated with a specific biochemical cascade to correlate receptor sequence with the type of enzyme cascade that it normally initiates. As regards the similarities between opsins from vertebrates, from *Drosophila* and also the β adrenergic receptor, an analysis based on minimal mutation distances suggests an evolutionary tree as illustrated in Figure 2. *Drosophila* opsin is not evolutionarily closer to vertebrate opsin than is the β adrenergic receptor to either opsin.[14] Thus, all may have diverged from a common gene for a membrane-bound receptor and there is no support from this analysis for the proposal that the rhodopsins of vertebrates and invertebrates initiate the same enzyme cascade. In view of the evidence, reviewed below, that the rhodopsins of invertebrates activate a phospholipase-C, rather than the cGMP phosphodiesterase activated by rod rhodopsins, the divergence of the opsin sequences is not unexpected.

A. EVIDENCE FOR LIGHT-ACTIVATED G PROTEINS IN MICROVILLAR PHOTORECEPTORS

It is now firmly established that the rhodopsin of vertebrate rod photoreceptors catalyzes the exchange of GTP for GDP on a G protein called transducin.[42] Transducin-GTP can then activate a cGMP-phosphodiesterase and reduce levels of the messenger, cGMP, that modulates the permeability of the rod's plasma membrane to cations.[48] Given that photoactivated rhodopsins from squid and octopus can cross react with transducin, it is attractive to propose that invertebrate rhodopsins also interact with a native G protein in the microvillar membrane. There is now substantial evidence for this proposition, which we review below.

1. Light-Activated GTP-ase Activity

After a receptor catalyzes the binding of GTP to the G protein and so activates it, hydrolysis of the bound GTP returns the G protein to the inactive state, in which the G protein has GDP bound to it. Thus, a light-activated GTP-ase activity would be expected in microvillar photoreceptors and has been reported in membrane preparations of octopus, squid, housefly (*Musca*), and blowfly (*Calliphora*) photoreceptors.[46,49-51] The light-activated GTP-ase is half-maximally stimulated by flashes that convert about 2% of rhodopsin to metarhodopsin. This sensitivity is less than that of the GTP-ase activity of rod photoreceptors from vertebrates, which is half-maximal when only 0.05% of the rhodopsin is bleached. Vandenberg and Montal[46] suggest that the lower sensitivity may be related to the lesser mobility of invertebrate rhodopsin in the microvillar membrane.

In all of the preparations of microvillar membranes, GTP-ase activity is maintained in the dark for tens of minutes after the initial flash of light, implying that some factor that normally turns off the transduction process is missing from these membrane preparations. This prolonged activity is reminiscent of the inability of the rod's cGMP-phosphodiesterase system to inactivate after a flash, if the photoreceptor disc membranes are washed of cytoplasmic components that normally inactivate metarhodopsin.[52] In rods, one mechanism that terminates the activation of transducin is the binding of a soluble 48 kDa protein to phosphorylated metarhodopsin.[53] A similar phosphorylation of invertebrate metarhodopsin occurs[20,21,54] and it has been suggested that phosphorylated metarhodopsin is unable to initiate phototransduction.[55-57] For fly photoreceptors,[50,51] the difference in the absorption spectra of metarhodopsin (red absorbing) and rhodopsin (blue absorbing)[8] enables the net reconversion of metarhodopsin back to rhodopsin by a red flash following an activating blue flash that caused a net conversion of rhodopsin to metarhodopsin. The reconversion quenches GTP-ase activity begun by an initial blue flash, demonstrating that the GTP-ase activity is associated with a metarhodopsin species that remains active long after a flash.

For preparations of purified cephalopod and blowfly microvillar membranes, the GTP-ase activity is not lost on washing in hypotonic solutions containing GTP, suggesting a tight association between G protein and membrane. In crude membrane preparations of housefly photoreceptors, however, much of the GTP-ase activity is lost upon hypotonic washing.[58] Whether this represents a species difference or whether a loosely bound fraction of G protein is lost during purification of the other preparations remains to be determined. It seems, though, that a fraction, at least, of the GTP-binding proteins are tightly bound to the microvillar membranes of several species, in contrast to the peripheral nature of transducin in the discs of vertebrate rods.

2. Light-Activated Binding of GTP-Analogs to Microvillar Membrane

As a prerequisite for light-activated GTP-ase activity, one would expect light to promote the binding of GTP to the G protein. One might also expect that light promotes the binding to the G protein of the hydrolysis-resistant analogs of GTP, GTPγS and GMP-PNP. Such binding of GTPγS and GMP-PNP has been observed in two of the preparations that show light-activated GTP-ase activity, those from squid[46,59] and housefly.[58] Binding saturates at a ratio of rhodopsin to bound GTPγS of between 10:1 and 100:1, indicating a minimum of one G protein per 10 to 100 rhodopsin molecules and, therefore, in a microvillus containing 1000 rhodopsin molecules, a minimum of 10 to 100 G proteins per microvillus.

B. THE IDENTIFICATION OF G PROTEINS IN MICROVILLAR MEMBRANES
1. Toxin-Catalyzed Labels

The alpha subunits of some GTP-binding proteins become ADP-ribosylated in the presence of cholera or pertussis toxins, enabling identification of the G protein after incubation with the toxin and radiolabeled NAD. In the case of photoreceptor G proteins, a modulation by light of the toxin-catalyzed labeling would imply the interaction of the labeled G protein with rhodopsin. One must bear in mind that events other than transduction, such as screening pigment movement and membrane turnover, may be regulated by G proteins and that the whole cell is liable to contain more than one target for the toxins. Thus, it is important to distinguish between studies of preparations enriched in microvillar membrane and those that use whole receptors or retinal homogenates. In studies of the latter type of preparation, light-independent ADP-ribosylation by cholera toxin of one protein of MW 43 kDa has been observed in homogenates of *Limulus* eyes.[60] Light-independent labeling of two proteins by pertussis toxin has been observed in homogenates of whole blowfly eyes (MW of substrates: 43 and 39 kDa respectively).[21] However, the significance of these substrates found in retinal homogenates for phototransduction and their localization within the eye is unknown.

All of the studies performed so far on preparations that are enriched in microvillar membrane show at least one site of *light-modulated* ADP-ribosylation in the presence of either toxin. However, the relative efficacy of the two toxins, the MW of the substrate and the effect of illumination vary with species. For instance, proteins of MW 44 and 41 kDa in, respectively, squid and blowfly microvillar membranes are labeled following treatment with cholera toxin,[21,46] but light enhances labeling of the former and diminishes labeling of the latter. A 41 kDa protein in octopus microvillar membranes can be labeled with pertussis toxin[61] labeling being diminished by light, but pertussis toxin is ineffective in labeling proteins in blowfly microvillar membranes.[21] How much of this variation is due to the differences in the methods of preparation and conditions of labeling and how much is due to genuine species variation remains to be determined, as does the significance of the opposite effects of light on cholera-toxin labeling in blowfly and squid photoreceptors. However, it is clear that all of the preparations enriched in microvillar membrane contain at least one protein showing light-modulated ADP-ribosylation catalyzed by cholera or pertussis toxin. Quantification of cholera-toxin catalyzed labeling of the 44 kDa protein in squid photoreceptor membranes[46] predicts a minimum of one G protein per 45 rhodopsin molecules, in good agreement with the density of G proteins estimated from GTPγS-binding (see above).

2. GTP Photoaffinity Labels

As an alternative to toxin-catalyzed labels, G proteins can be identified using a radioactively labeled photo-affinity analog of GTP, azidoanilido GTP. This technique has recently been applied to crude membrane preparations of housefly photoreceptors.[58] After a blue light that causes a net conversion of rhodopsin to metarhodopsin, the membranes were absorbed onto a nitrocellulose filter and washed. The photoaffinity analog was then activated by an intense UV light. Proteins with MW 41 and 39 kDa were labeled. Preillimunation with a red light, which did not cause a net conversion of rhodopsin to metarhodopsin, resulted in the labeling of only the 39 kDa proteins. Thus, a 41 kDa protein that is stimulated by metarhodopsin to bind GTP and a 39 kDa light-independent G protein exist in this crude photoreceptor membrane preparation. These proteins may be homologous with the 41 kDa light-modulated cholera-toxin substrate that is associated with purified blowfly microvillar membranes and the 39 kDa light-independent pertussis toxin substrate found in crude blowfly photoreceptor membrane.[21]

A third method for identifying possible GTP-binding proteins, immunological cross-reaction with antibodies raised against the alpha subunit of transducin, has not so far been successful. Presumably the structures of the alpha subunits of invertebrate and vertebrate photoreceptor GTP-binding proteins are too dissimilar, although a limited homology may exist.[45] An antibody raised against the β subunit of transducin does cross-react with a 36 kDa polypeptide in octopus photoreceptors[61] and there is evidence that the β subunit is highly conserved among known G proteins.[62,63]

In summary, the microvillar membranes of cephalopod and fly photoreceptors show the

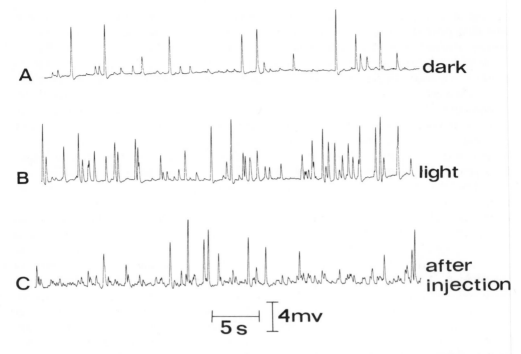

FIGURE 3.   Excitation of a *Limulus* ventral photoreceptor by GTPγS. (A) In darkness, the membrane potential of the cell exhibits spontaneous discrete waves of depolarization. (B) The frequency of the discrete waves is increased by light. Each photon absorbed by rhodopsin initiates a single new discrete wave. (C) Injection of GTPγS in darkness also increases the rate of occurrence of the discrete events, particularly a class of events with a smaller amplitude distribution than those evoked by light. (Adapted from Fein, A. and Corson, D. W., *Science*, 212, 555, 1981.)

light-dependent GTP-ase activity, GTP-binding, and *light-modulated* toxin-catalyzed labeling expected if photoactivated rhodopsin catalyzes the exchange of GTP for GDP on a GTP-binding protein. There appear to be many copies of G protein per microvillus, but it is not possible, yet, to assign a particular protein as the unique source of the GTP-ase activity. Also, the possibility that more than one species of light-activated G protein exists in the microvillus cannot be ruled out. Further understanding must await the reconsitution of membranes containing rhodopsin and purified GTP-binding protein. Progress towards this goal is limited by the tight association in most species of the G protein to the microvillar membrane.

III. PHYSIOLOGY OF MICROVILLAR PHOTORECEPTORS

A. PHYSIOLOGICAL EVIDENCE FOR THE INVOLVEMENT OF G PROTEINS IN PHOTO-TRANSDUCTION

Experiments in which the photoreceptors of invertebrates have been exposed to agents that activate or inhibit GTP-binding proteins have provided evidence for the involvement of GTP-binding proteins in phototransduction. Fluoride activates transducin in darkness[64] and is also able to activate the light-sensitive conductance of invertebrate photoreceptors. Exposure of *Limulus* ventral[65] and insect photoreceptors[66,67] to fluoride produces a depolarization composed of discrete events that are similar to, but smaller than, those produced by single photons. A more specific agent than fluoride which also activates transducin[68] and the GTP-binding protein that modulates adenylate cyclase activity[69] is GTPγS, a nonhydrolyzable analog of GTP. GTPγS has similar effects to fluoride on *Limulus* and fly photoreceptors (Figure 3).[67,70,71] The guanine nucleotide binding site of the G protein can also be probed using an analog of GDP, GDPβS. Intracellular injection of GDPβS blocks excitation and adaptation of *Limulus* ventral photoreceptors by light.[72] This blockade may result from binding of GDPβS to the guanine nucleotide binding site of the G protein, preventing the activation of the G protein by light-induced exchange of GTP for GDP. This is the proposed mechanism for the inhibition of hormone-activated adenylate cyclase by GDPβS.[73]

The above experiments lend support to the involvement of a GTP-binding protein in transduction but give no clue as to the enzyme that is activated by the G protein which, in turn, initiates excitation and adaptation of the photoreceptor. Neither do they link this light-activated GTP-binding protein to the release of calcium. Before considering this problem, we first briefly review the final consequences of the visual cascade in microvillar photoreceptors — the opening of ionic channels in the plasma membrane and the release of calcium from intracellular stores.

B. RELEASE OF CALCIUM FROM INTERNAL STORES BY LIGHT

A light-induced rise in Ca_i has been found in several microvillar photoreceptors, including those of barnacle,[1] bee,[74] the marine mollusc *Hermissenda*,[75] and the horseshoe crab *Limulus*.[1,76,78] The most extensive studies have been performed on the giant ventral photoreceptors of *Limulus*. These photoreceptors are clearly segmented into two lobes (Figure 4A), only one of which (the rhabdomeral of R-lobe) bears microvilli and is therefore light sensitive.[79,80] Measurements in *Limulus* ventral photoreceptors using a variety of techniques show that bright flashes of light rapidly increase Ca_i.[1,76,78] The rise in calcium is not uniform within the cell,[81] but it is largely confined to the light-sensitive R-lobe of the cell (Figure 4B).[77,82] Indeed, if illumination is confined to a 10 to 20 μm diameter spot within the 50 to 100 μm diameter R-lobe, then the rise in Ca_i is further confined to that spot.[83] Following a brief flash of light, the timecourse of the rise in Ca_i is very rapid, reaching its peak amplitude of approximately 40 μM in less than 300 ms and declining to half-maximal amplitude within 2 s (Figure 4C). If illumination is maintained for several seconds, Ca_i falls from this peak to a sustained plateau level of a few micromolar.[77]

The light-induced rise in Ca_i is diminished by less than 50% when *Limulus* ventral photoreceptors are soaked, in darkness, with seawater containing little or no calcium, suggesting that a substantial fraction of the large transient rise in Ca_i that follows a light flash is due to the release of calcium from internal stores rather than entrance of calcium from the bathing medium.[1,77,84] In other microvillar photoreceptors, such as those of the barnacle, the fraction of calcium entering the receptor from the outside via light or voltage-activated calcium-permeable ionic channels may be larger.[1] Also, it is probable in *Limulus* ventral photoreceptors that the elevation in Ca_i that follows the initial transient during a long step of light may be sustained, in part, by calcium influx.

The identification of the stores of calcium that are released by light in *Limulus* photoreceptors is aided by the specific localization of calcium release to the R-lobe, which uniquely bears microvilli. As shown schematically in Figure 4, close to the bases of the microvilli lie cisternae of smooth ER, the submicrovillar cisternae (SMC).[79,85] These cisternae are ubiquitous features of microvillar photoreceptors and their morphology has been extensively studied (see review by Whittle[86]). The SMC have been shown to use ATP to actively accumulate calcium[33,87-89] and so may function both as a source and as a sink for the calcium released by light. The cisternae are ideally localized very close to the bases of the microvilli. Measurements in fly and leech photoreceptors reveal gaps of only 9 ± 2 and 24 ± 9 nm, respectively,[33,89] but there is, nevertheless, the need for an intracellular messenger to cross the gap between photoactivated rhodopsin molecules located in the plasma membrane and the nearby cisternae.[90] Evidence that inositol triphosphate[91] fulfills this messenger role will be reviewed below.

Two features of the rise in Ca_i in *Limulus* ventral photoreceptors are of interest in

A

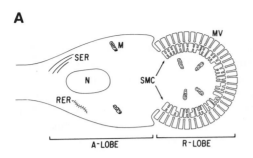

B

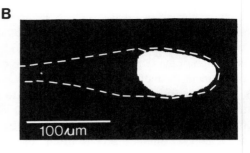

100 μm

C

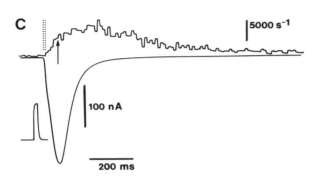

5000 s⁻¹

100 nA

200 ms

FIGURE 4.   Characteristics of the light-induced rise in Ca$_i$ in a *Limulus* ventral photoreceptor. (A) Diagram of a cross section through a *Limulus* ventral photoreceptor, showing the morphology of the A- and R-lobes of the cell. The A-lobe contains the nucleus (N), rough endoplasmic reticulum (RER), smooth endoplasmic reticulum (SER), and mitochondria (M). The R-lobe contains microvilli (MV), subrhabdomeral cisternae of smooth ER (SMR), and mitochondria. (B) Light-induced aequorin luminescence recorded with an image intensifier from a *Limulus* ventral photoreceptor following the delivery of a 10 ms white light flash that uniformly illuminated the entire cell. The dashed line outlines the photoreceptor cell body. The aequorin luminescence, which signals the light-induced rise in Ca$_i$, is confirmed to the distal end of the cell body which comprises the light-sensitive R-lobe, as shown in (A). For experimental details see Payne and Fein.[82] (C) A comparison of the time-course of light-induced aequorin luminescence and photocurrent recorded from a *Limulus* ventral photoreceptor after a flash of light (light monitor is shown at bottom left). The photocurrent (lower smooth trace) and the accompanying aequorin luminescence (upper noisy trace) were elicited by a 20 ms flash. The arrow shows the time of the peak photocurrent which preceeds the peak aequorin luminescence. Dotted lines show the period within which the photomultiplier registered the first counts of luminescence prior to the output of the photon counter, demonstrating the similar latent periods of photocurrent and aequorin luminescence.

considering the mechanism of release. First, the large magnitude of the rise in Ca$_i$ implies that a biochemical amplifier must mediate the release of several thousand calcium ions in response to one photoisomerization.[77] For *Limulus* ventral photoreceptors, a flash that isomerizes approximately 10^5 rhodopsin molecules causes a peak rise in Ca$_i$ of 40 μM in a microvillar lobe having a typical volume of 100 pl. This implies a release of at least 10^4 calcium ions per single photoactivated rhodopsin molecule. A second consideration is the rapidity of the calcium-release (see Figure 4C). Ca$_i$ begins to rise within 50 ms of a flash of light, implying that the biochemical elements in the chain of amplification that releases calcium must be physically close together. This is achieved, probably, by the tight apposition of the SMC to the bases of the microvilli and by the high concentration of the biochemical elements of the visual cascade in the microvillus. Thus, the spatial localization of the rise in calcium (see Figure 4B) is probably a necessary consequence of the need for a rapidly activating system.

C. THE ROLE OF CALCIUM IONS IN THE RESPONSE TO LIGHT

The end result of excitation of microvillar photoreceptors is an electrical depolarization of the cell's plasma membrane. The role of the light-induced rise in Ca_i in mediating and modulating this depolarization is controversial and a detailed discussion is beyond the scope of this chapter. The following is a brief survey of the current state of our understanding.

Illumination of microvillar photoreceptors specifically increases the cationic permeability of the plasma membrane in the region of the cell that contains microvilli, resulting in a flow of current into the cell.[92-95] This light-activated current (photocurrent), carried mainly by sodium ions[96] through light-activated ionic channels,[97] depolarizes the membrane of the entire cell body. The photocurrent can be monitored by impaling the cell with two glass capillary mcroelectrodes so as to examine the transmembrane currents under voltage-clamp. Simultaneously, the rise in Ca_i following a flash can be conveniently monitored by microinjecting a *Limulus* ventral photoreceptor with the Ca-sensitive photoprotein aequorin.[98] The rise in Ca_i is monitored from the Ca-sensitive luminescence of aequorin. Figure 4C shows that the photocurrent and the aequorin luminescence both begin to rise after a latent period of 50 ms following a flash of light. The photocurrent peaks before the aequorin luminescence, at about 200 ms after the flash. Thus, the rise in Ca_i accompanies both the rising and falling phases of the photocurrent, but its peak lags behind that of the photocurrent.

Prolonged or bright illumination adapts the photoreceptor by rapidly reducing the ability of light to activate the photocurrent. The consequent rapid (≤ 2 s) decline of the depolarization during a bright step of light avoids saturation of the voltage-response of the photoreceptor. Thus, there are two processes initiated by rhodopsin: (1) the generation of the photocurrent, *excitation*, and (2) the reduction by *adaptation* of the ability of light to excite the cell. Calcium has been proposed to play a role in both processes.

The evidence in favor of calcium as a messenger for adaptation is strong. The essential feature of adaptation, the reduction of the sensitivity of the photocurrent to light, can be mimicked by intracellular injection of calcium (Figure 5A)[99] and opposed by the injection of calcium chelators.[100] In addition, the rise in calcium (Figure 4C) is rapid enough to mediate the 1 to 2 s timecourse of the onset of adaptation. Thus, calcium released by light is an important factor in mediating light adaptation.

The role of the rise in Ca_i in mediating excitation by light is controversial. Pressure injection of calcium[101] or the calcium-releasing agent $InsP_3$ (see below) into the rhabdomeral lobe of *Limulus* ventral photoreceptors opens a conductance with similar properties to that opened by light, showing that a rise in Ca_i is a sufficient signal to excite the photoreceptor under some conditions (Figure 5A and B). However, the injection of calcium chelators does not diminish excitation by light, although it slows the process down.[100] Neither does the rising edge of the light-induced rise in Ca_i, as indicated by aequorin luminescence, seem to preceed the generation of the photocurrent (Figure 4C). Calcium therefore is able to excite the photoreceptor, but it may not be the only pathway by which light can open ionic channels. Other pathways for the direct activation of the light-sensitive ionic channels, such as a light-induced rise in the concentration of the messenger cGMP may also exist.[102,103] Unfortunately, lack of knowledge of the degree to which calcium chelators and indicators such as aequorin can rapidly mix with the calcium that light releases into the space between microvilli and subrhabdomeral cisternae means that we cannot definitely rule out the possibility that calcium is a messenger of excitation as well as adaptation.

In neither the case of adaptation nor excitation do we know the chemical nature of the sites at which calcium acts. Thus, we do not know whether calcium excites the cell by opening ionic channels directly or by modulating the level of another messenger, which in turn opens the channels.

D. INOSITOL 1,4,5 TRIPHOSPHATE MEDIATES CALCIUM-RELEASE

Inositol 1,4,5 triphosphate, $InsP_3$ has been proposed as the messenger that is released by rhodopsin from the plasma membrane and which can then cross the gap to the SMC so

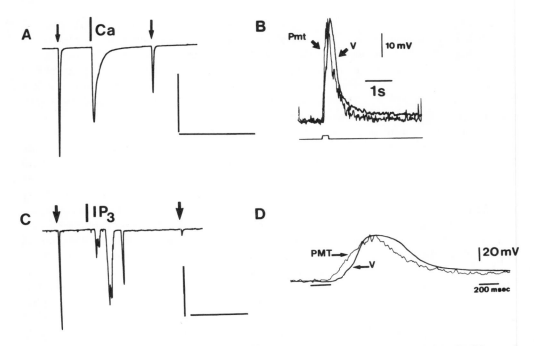

FIGURE 5.   Excitation and adaptation of a *Limulus* ventral photoreceptor by pressure injection of calcium and InsP$_3$. (A) Excitation and adaptation by calcium. The trace shows the photoreceptor's transmembrane current, recorded under voltage clamp. Diffuse light flashes were delivered at the times indicated by arrows. A brief pulse of 2 m*M* calcium was delivered by pressure through a micropipette into the R-lobe at the time indicated by the bar above the trace. Like light, the injection excited the cell by eliciting a transient inward current. The response to a subsequent light flash was diminished by the calcium injection, mimicking light adaptation. Recovery of the response to light was complete 40 s later. The vertical scale bar represents 10 nA; the horizontal 4 s. (Adapted from Payne, R., Corson, D. W., and Fein, A., *J. Gen. Physiol.*, 88, 107, 1986.) (B) The relationship between aequorin luminescence (PMT) and the depolarization (V) caused by the calcium-activated current. The trace representing aequorin luminescence is arbitrarily scaled to the same peak amplitude as the trace representing depolarization. The depolarization closely follows the rise in Ca$_i$ as indicated by aequorin luminescence. The bottom trace indicates the time of the pressure injection of 2 m*M* calcium into the R-lobe. (Adapted from Payne, R., Corson, D. W., and Fein, A., *J. Gen. Physiol.*, 88, 107, 1986. (C) Excitation and adaptation by InsP$_3$. The trace shows the photoreceptor's transmembrane current, recorded under voltage clamp, following injection of 100 μ*M* InsP$_3$ (bars) and 10 ms light flashers (arrows). Both light and InsP$_3$ injection excited the cell by eliciting transient inward currents. InsP$_3$ also mimicked light adaptation insofar as the response to a second light flash was diminished by the injection of InsP$_3$. Recovery of the responses to light and InsP$_3$ was complete within 5 min. The cell was voltage clamped to its dark-adapted resting potential throughout the experiment. The vertical scale bar represents 5 nA; the horizontal, 10 s. (Adapted from Payne, R., Corson, D. W., Fein, A., and Berridge, M. J., *J. Gen. Physiol.*, 88, 127, 1986.) (D) Aequorin luminescence and membrane depolarization elicited by injection of InsP$_3$ into a ventral photoreceptor. Injection of 1 m*M* InsP$_3$ at the time indicated by the bar beneath the traces elicited both depolarization of the cell membrane (trace labeled V) and aequorin luminescence (trace labeled PMT) which indicated a rise in Ca$_i$. The PMT trace is arbitrarily normalized to the peak of the voltage trace. (Adapted from Payne, R., Corson, D. W., Fein, A., and Berridge, M. J., *J. Gen. Physiol.*, 88, 127, 1986.)

as to mobilize calcium stores.[104,105] In support of this hypothesis, injection of InsP$_3$ into *Limulus* ventral photoreceptors releases calcium from internal stores, resulting in a rise in intracellular calcium, similar in amplitude and timecourse to that produced by a flash of light (Figure 5D).[106,107] Like light, InsP$_3$ specifically releases calcium from stores in the microvillar R-lobe of the ventral photoreceptors, supporting the idea that the SMC are a unique target for the action of InsP$_3$.[82]

As might be expected from the above discussion of the actions of calcium, pressure injection of InsP$_3$ into *Limulus* ventral photoreceptors both excites and adapts the photoreceptor (Figure 5C).[104,105] Injections of 2 to 100 μ*M* InsP$_3$ produce bursts of depolarization

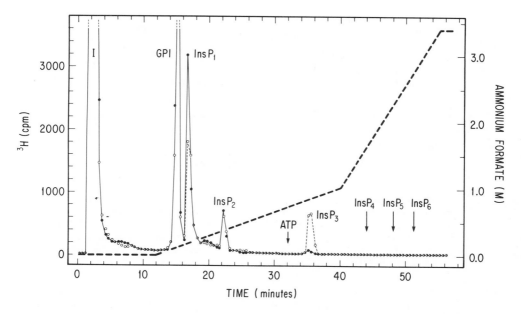

FIGURE 6.   Light-induced production of InsP₃ in squid retina. Separation by HPLC of the inositol phosphates contained in the water extract from an unstimulated half-eyecup (●) and its light-stimulated pair (○) that was quenched 3 s after a flash. (▪) indicates overlapping data points. The left ordinate indicates the radioactivity in eluted fractions. The right ordinate refers to the eluting solvent strength shown by the thick broken line. Only Ins(1,4,5)P₃ showed a significant light-induced increase. (From Szuts, E. Z., Wood, S. F., Reid, M. A., and Fein, A., *Biochem. J.*, 240, 929, 1986. With permission.)

of amplitude 5 to 50 mV and each burst of depolarization is associated with a concommitant rise in Ca$_i$ (Figure 5D).[108] Prior injection of calcium chelators abolishes the depolarization caused by injections of InsP₃,[107,109] suggesting that InsP₃ acts solely by releasing calcium and that it has no direct action on the light-sensitive conductance. Recently, InsP₃ has been injected into receptors of the lateral compound eye of *Limulus* creating bursts of depolarization similar to those seen in ventral photoreceptors.[110] InsP₃ has also been introduced into photoreceptors of the housefly compound eye, producing a sustained depolarization made up from the summation of many small events similar to those elicited by single photons.[58]

As well as exciting *Limulus* ventral photoreceptors, injection of InsP₃ reversibly adapts subsequent responses both to diffuse light flashes (Figure 5C)[104,105] and to further injections of InsP₃.[105] The ability of InsP₃ to adapt ventral photoreceptors is also abolished by prior injection of calcium chelators, suggesting that it, too, arises solely from the release of calcium. The natural breakdown product of Ins(1,4,5)P₃, Ins(1,4)P₂, is at least ten times less effective than Ins(1,4,5)P₃ in exciting and adapting the photoreceptor,[104,105] consistent with its lesser ability to release calcium from stores in other cells.

E. EVIDENCE FOR THE LIGHT-INDUCED PRODUCTION OF Ins(1,4,5)P₃

Light-induced increases in InsP₃ content have been reported for *Limulus* ventral eye[104] and squid retinae (Figure 6).[111,112] In each case, the tissue was incubated with ³H-inositol so as to form ³H-phosphoinositide in the plasma membrane. Breakdown of ³H-phospho-inositides on illumination produces detectable ³H-inositol phosphates. Light-induced production of similarly labeled InsP₃ from membrane preparations of squid and housefly eyes has also been observed.[58,113] These experiments expanded on earlier reports of phospho-inositide turnover in cephalopod retinae.[54,114] Production of InsPi₃ content are observed in squid retinae within 200 ms.[111] Thus, production of InsP₃ in squid eyes occurs over the same timescale as release of calcium in *Limulus* ventral photoreceptors. Light-induced decreases

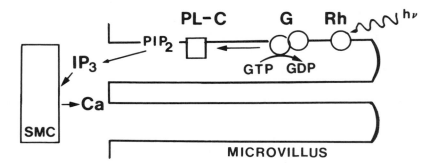

FIGURE 7.   Cascade proposed by Fein[72] for the coupling of photoactivated rhodopsin to the release of inositol trisphosphate. After absorption of a photon, photoactivated rhodopsin (Rh) catalyzes the exchange of GTP for GDP on a G-protein (G). The G protein, with GTP bound, activates a phospholipase-C which cleaves inositol trisphosphate from phosphatidylinositol bisphosphate (PIP$_2$). Inositol trisphosphate then releases calcium from the submicrovillar cisternae (SMC).

in the lipid precursor of InsP$_3$, Phosphatidylinositol 1,4 bisphosphate (PIP$_2$) have been observed in squid and *Limulus* preparations[54,104,112] suggesting that the InsP$_3$ is produced, as in other cells, by the activation of a phospholipase-C which hydrolyzes PIP$_2$. This has been directly confirmed in membrane preparations of squid photoreceptors by observing light-activated hydrolysis of exogenous ^{3}HPIP$_2$.[115]

In most of these preparations, the content of InsP$_2$ and InsP$_1$ also rises after a light flash. These inositol phosphates could arise from hydrolysis of either the corresponding parent lipids by phospholipase-C or by sequential hydrolysis of InsP$_3$. The specificity of the light-activated phospholipase-C for PIP$_2$ over PIP or PI has not been directly determined, although in *Limulus* and squid preparations the content of PIP$_2$ falls after illumination while that of PIP is not significantly reduced.[104,112] This suggests that the initial action of light is the hydrolysis of PIP$_2$ to produce InsP$_3$, which is then degraded by specific phophatases to InsP$_2$ and InsP$_1$.[116] This interpretation is supported by the observations that production of InsP$_3$ in outer segments of squid retinae after a flash precedes that of InsP$_2$ (S.F. Wood personal communication). The production of InsP$_2$ in the latter preparation and in a preparation of housefly photoreceptor membranes[58] can be inhibited by the addition of 2,3 diphosphoglyceric acid, which inhibits the phosphatase that hydrolyzes InsP$_3$.[117]

Although some of the Ins(1,4,5)P$_3$ is degraded to the less active Ins(1,4)P$_2$, it is possible that in some preparations, some Ins(1,4,5)P$_3$ is phosphorylated to Ins(1,3,4,5)P$_4$. Illumination of *Limulus* ventral eyes[118] (but not squid eyes[111,112]) results in the production of inositol 1,3,4 triphosphate in addition to Ins(1,4,5)P$_3$. Ins(1,3,4)P$_3$ is thought to be formed by phosphorylation of Ins(1,4,5)P$_3$ to Ins(1,3,4,5) tetrakisphosphate followed by dephosphorylation to Ins(1,3,4)P$_3$.[119,120] However, inositol tetrakisphosphate has not so far been detected in photoreceptors of invertebrates and the role of inositol 1,3,4 triphosphate in excitation and adaptation of *Limulus* ventral photoreceptors remains to be elucidated.

F. THE CONTROL OF PHOSPHOLIPASE-C BY RHODOPSIN

Several groups are now trying to obtain evidence that rhodopsin is coupled to the activation of phospholipase-C via a G-protein. Figure 7 summarizes the proposed model of the coupling.[72] The effects on inositol phosphate production of fluoride, GTPγS and GDPβS, have been examined. These agents are known to activate or inhibit G-binding proteins and they also mimic or inhibit the electrical response to light (see above). Preparations of housefly and squid photoreceptor membranes have so far been examined. Devary et al.[58] report that GTPγS enhanced the accumulation of InsP$_2$ following a light flash. Rapid hydrolysis of

InsP$_3$ to InsP$_2$ in this preparation prevented the analysis of InsP$_3$ content. Fluoride caused a similar accumulation of InsP$_2$ in darkness, while addition of GDPβS completely inhibited the accumulation of InsP$_2$ following illumination or treatment with fluoride. Fluoride has also been applied to a preparation of squid photoreceptor membranes.[121] Increased content of InsP$_3$ and accumulation of InsP$_2$ were observed after incubation with fluoride in darkness. A correspondingly decreased PIP$_2$ content suggests that, like light, fluoride activates phospholipase-C.

A more direct demonstration of GTP dependence of InsP$_3$ production is prevented in preparations of crude photoreceptor membranes by the presence of contaminating micromolar GTP, which is sufficient to sustain the GTP-ase activity even if no GTP is added. Only a small (20%) increase in light-induced InsP$_3$ production is observed on addition of 1 mM GTP to squid membrane preparations.[121] To overcome this problem, Baer and Saibil[115] have developed a purified preparation of squid photoreceptors that is washed to remove endogenous GTP. Radioactively labeled PIP$_2$ is added as a substrate for the light-activated phospholipase-C. At a calcium concentration of 1 μM, light-activated production of InsP$_3$ from the exogenous substrate showed an absolute requirement for GTP. At lower calcium concentrations light production of InsP$_3$ was superimposed upon a light- and GTP-independent background. The relevance of the calcium-dependence of InsP$_3$-production in this purified preparation to more physiological preparations remains to be determined, as does the K$_m$ for GTP. Thus, these experiments provide evidence that the production of InsP$_3$ is mediated through the activation of a G protein. The identity of the protein and its relationship to the G protein that is responsible for light-activated GTP-ase and GTP-binding is a target for future research.

IV. SUMMARY

The experiments reviewed in this chapter provide evidence that light activates a phospholipase-C in photoreceptors of invertebrates via the activation by rhodopsin of G-binding protein. The resulting production of InsP$_3$ then releases calcium from smooth endoplasmic reticulum beneath the microvilli. The result is a powerful biological amplifier that releases many thousand calcium ions in response to the absorption of a single photon. The evidence consists, primarily, of qualitative demonstrations that one link in the chain is a sufficient activator of the next. Thus, light activates a G-protein, light releases InsP$_3$, and InsP$_3$ releases calcium so as to adapt and possibly excite the receptor. One weakness of the experiments lies in the lack of quantitative proof that the concentration of each mediator of the cascade is sufficient to trigger the next step. A second weakness is our inability to prove that the proposed cascade is the *only* possible way in which information flows in the transduction process. The complete blockade of excitation and adaptation of *Limulus* ventral photoreceptors by GDPβS suggests that activation of a G-protein is the *only* means by which rhodopsin initiates visual cascade. However, it is possible that messengers other than InsP$_3$ could release calcium or (more likely) directly open the light-sensitive conductance. Light-induced elevations of cGMP have been proposed as an additional mechanism for activating the light-induced conductance in *Limulus* and squid photoreceptors.[102-103] The development of new tools, such as specific blockers of InsP$_3$-induced calcium-release or phospholipase-C will probe the sufficiency of our model of calcium release to explain excitation and adaptation of the photoreceptors by light.

REFERENCES

1. **Brown, J. E. and Blinks, J. R.**, Changes in intracellular free calcium during illumination of invertebrate photoreceptors. Detected with aequorin, *J. Gen. Physiol.*, 64, 643, 1974.
2. **Yau, K.-W. and Nakatani, K.**, Light-induced reduction of cytoplasmic free calcium in retinal rod outer segments, *Nature (London)*, 313, 579, 1985.
3. **MacNaughton, P. A., Cevetto, L., and Nunn, B. J.**, Measurement of the intracellular free calcium in concentration in salamander rods, *Nature (London)*, 332, 261, 1986.
4. **Ratto, G. M., Payne, R., Owen, W. G. and Tsien, R. Y.**, The concentration of cytosolic free calcium in vertebrate rod outer segments measured with Fura-2, *J. Neurosci.*, 8, 3240, 1988.
5. **Kirschfeld, K.**, Activation of visual pigment: chromophore structure and function, in *The Molecular Mechanism of Photoreception*, Dahlem Konferenzen 1986, Stieve, H., Ed., Springer-Verlag, Berlin, 1986, 31.
6. **Findlay, J. B. C.**, The biosynthetic, functional and evolutionary implications of the structure of rhodopsin, in *The Molecular Mechanism of Photoreception*, Dahlem Konferenzen 1986, Stieve, H., Ed., Springer-Verlag, Berlin, 1986, 11.
7. **Applebury, M. L. and Hargrave, P. A.**, Molecular biology of visual pigments, *Vision. Res.*, 26, 1881, 1986.
8. **Hamdorff, K.**, The physiology of invertebrate visual pigments, in *The Handbook of Sensory Physiology*, Vol. 7/6A, Autrum, H., Ed., Springer-Verlag, Berlin, 1979, 145.
9. **Harris, W. A., Stark, W. S., and Walker, J. A.**, Genetic dissection of the photoreceptor system in the compound eye of *Drosophila melanogaster, J. Physiol. (London)*, 256, 415, 1976.
10. **Zuker, C. S., Cowman, A. F., and Rubin, G. M.**, Isolation and structure of a rhodopsin gene from *D. melanogaster, Cell*, 40, 841, 1985.
11. **O'Tousa, J. E., Baehr, W., Martin, R. L., Hirsch, W., Pak, W. L., and Applebury, M. L.**, The *Drosophila* ninaE gene encodes an opsin, *Cell*, 40, 839, 1985.
12. **Cowman, A. F., Zuker, C. S., and Rubin, G. M.**, An opsin gene expressed in only one photoreceptor cell type of the Drosophila eye, *Cell*, 44, 705, 1986.
13. **Montell, C., Jones, K., Zuker, C., and Rubin, G.**, A second opsin gene expressed in the ultra-violet-sensitive R7 photoreceptor cells of Drosophila melanogaster, *J. Neurosci.*, 7, 1558, 1987.
14. **Zuker, C. S., Montell, C., Jones, K., Laverty, T., and Rubin, G. M.**, A rhodopsin gene expressed in photoreceptor cell R7 of the Drosophila eye: homologies with other signal-transducing molecules, *J. Neurosci.*, 7, 1550, 1987.
15. **Hardie, R. C.**, The photoreceptor array of the dipteran retina, *Trends. Neurosci.*, 9, 419, 1986.
16. **Ovchinnikov, Y. A., Abdulaev, N. G., Feigina, M. Y., Artomonov, I. D., Zolotarev, A. S., Moroshnikov, A. I., Martynow, V. I., Kostina, M. B., Kudelin, A. G., and Bogachuk, A. S.**, The complete amino acid sequence of visual rhodopsin, *Biog. Khim.*, 18, 1424, 1982.
17. **Vogt, K. and Kirschfeld, K.**, Chemical identity of the chromophore of fly visual pigment, *Z. Naturforsch.*, 71, 211, 1984.
18. **Seki, T., Fuujishita, S., Ito, M., Nozomi, M., Kobayashi, C., and Tsukida, K.**, A fly, *Drosophila melanogaster* forms 11-cis-3hydroxyretinal in the dark, *Vision Res.*, 26, 255, 1986.
19. **Kruizinga, B., Kamman, R. L., and Stavenga, D. G.**, Laser-induced visual pigment conversions in fly photoreceptor. Measured *in vivo, Biophys. Struct. Mech.*, 9, 299, 1983.
20. **Paulsen, R. and Hoppe, I.**, Light-activated phosphorylation of cephalopod rhodopsin, *FEBS Lett.*, 96, 55, 1978.
21. **Bentrop, J. and Paulsen, R.**, Light-modulated ADP-ribosylation, protein phosphorylation and protein binding in isolated fly photoreceptor membranes, *Eur. J. Biochem.*, 161, 61, 1986.
22. **Thomspon, P. and Findlay, J. B. C.**, Phosphorylation of ovine rhodopsin, identification of the phosphorylated sites, *Biochem. J.*, 220, 773, 1984.
23. **Eguchi, E. and Waterman, T. H.**, in *The Functional Organisation of the Compound Eye*, Bernhard, C. G., Ed., Pergamon Press, Oxford, 1966, 105.
24. **Shaw, S. R.**, Sense cell structure and interspecies comparisons of polarized light absorption in arthropod compound eyes, *Vis. Res.*, 9, 1031, 1969.
25. **Anderson, R. E., Benolken, R. M., Kelleher, P. A., Maude, M. B., and Wiegard, R. D.**, Chemistry of photoreceptor membrane preparations from squid retinae, *Biochim. Biophys. Acta*, 510, 316, 1978.
26. **Akino, T. and Tsuda, M.**, Characteristics of phospholipids in microvillar membrane of octopus photoreceptor cells, *Biochim. Biophys. Acta*, 556, 61, 1979.
27. **Paulsen, R., Zinkler, D., and Delmelle, M.**, Architecture and dynamics of microvillar photoreceptor membranes of cephalopods, *Exp. Eye Res.*, 36, 47, 1983.
28. **Saibil, H. and Hewat, E.**, Ordered transmembrane and extracellular structure in squid photoreceptor microvilli, *J. Cell. Biol.*, 105, 19, 1987.

29. **Varela, F. G. and Porter, K. R.,** Fine structure of the visual system of the honeybee *Apis mellifera*. I. The retina, *J. Ultrastruct. Res.,* 29, 236, 1969.

30. **Blest, A. D., Stowe, S., and Eddey, W.,** A labile calcium-dependent cytoskeleton in rhabdomeral microvilli of blowflies, *Cell. Tissue Res.,* 223, 553, 1982.

31. **Saibil, H. R.,** An ordered membrane cytoskeleton network in squid photoreceptor microvilli, *J. Mol. Biol.,* 158, 435, 1982.

32. **DeCouet, H. G., Stowe, S., and Blest, A. D.,** Membrane-associated actin in the rhabdomeral microvilli of crayfish photoreceptors, *J. Cell Biol.,* 98, 834, 1984.

33. **Walz, B.,** ATP-dependent calcium-uptake by smooth endoplasmic reticulum in an invertebrate photoreceptor cell. An ultrastructural, cytochemical and X-ray microanalytical study, *Eur. J. Cell Biol.,* 20, 83, 1979.

34. **Blest, A. D., DeCouet, H. G., and Sigmund, C.,** The microvillar cytoskeletion of leech photoreceptors: a stable bundle of actin microfilaments, *Cell. Tissue Res.,* 234, 9, 1983.

35. **Foster, M. C.,** Solution of the diffusion equation for membranes containing microvilli and lateral diffusion of rhodopsin and lipid analogues in squid photoreceptor membranes, *Fed. Proc.,* 39, 2067, 1980.

36. **Goldsmith, T. H. and Wehner, R.,** Restrictions on rotational and translational diffusion of pigment in membranes of a rhabdomeric photoreceptor, *J. Gen. Physiol.,* 70, 453, 1977.

37. **Liebman, P. A. and Entine, G.,** Lateral diffusion of visual pigment in photoreceptor disk membranes, *Science,* 185, 457, 1974.

38. **Poo, M. and Cone, R. A.,** Lateral diffusion of rhodopsin in photoreceptor membranes, *Nature (London),* 247, 438, 1974.

39. **Kubo, T., Fukuda, K., Mikami, A., Maeda, A., Takahashi, H., Mishina, M., Haga, T., Haga, K., Ichiyama, A., Kangawa, K., Kojima, M., Matsuo, H., Hirose, T., and Numa, S.,** Cloning, sequence and expression of complementary DNA encoding the muscarinic acetylcholine receptor, *Nature,* 323, 411, 1986.

40. **Dixon, R., Kobilka, B., Strader, B., Benovic, J., Dohlman, H., Frielle, T., Bolanowski, M., Bennet, C., Rands, E., Diehl, R., Mumford, R., Slater, E., Sigal I., Caron, M. G., Lefkowitz, R. J., and Strader, C. D.,** Cloning of the gene and cDNA for the mammalian β adrenergic receptor and homology with rhodopsin, *Nature,* 321, 75, 1986.

41. **Hall, Z. W.,** Three of a kind: the β-adrenergic receptor, the muscarinic acetylcholine receptor and rhodopsin, *Trends Neurosci.,* 10, 99, 1987.

42. **Stryer, L.,** The cGMP cascade of vision, *Annu. Rev. Neurosci.,* 9, 87, 1986.

43. **Levitski, A.,** β-adrenergic receptors and their mode of coupling to adenylate cyclase, *Phys. Rev.,* 66, 819, 1986.

44. **Ebrey, T. G., Tsuda, M., Sassanrath, G., West, J. L., and Waddell, W. H.,** Light-activation of bovine rod phosphodiesterase by non-physiological visual pigments, *FEBS Lett.,* 116, 217, 1980.

45. **Saibil, H. R. and Michel Villaz, M.,** Squid rhodopsin and GTP-binding protein cross-react with vertebrate photoreceptor systems, *Proc. Natl. Acad. Sci. U.S.A.,* 81, 5111, 1984.

46. **Vandenberg, C. A. and Montal, M.,** Light-regulated biochemical events in invertebrate photoreceptors. I. Light-activated guanosine-triphosphate, guanine nucleotide binding, and cholera toxin labeling of squid photoreceptor membranes, *Biochemistry,* 23, 2339, 1984.

47. **Bitensky, N. W., Wheeler, M. A., Rasenick, M. M., Yamazaki, A., Stein, P., Halliday, K. R., and Wheeler, G. L.,** Functional exchange of components between light-activated photoreceptor phosphodiesterase and hormone-activated adenylate cyclase systems, *Proc. Natl. Acad. Sci., U.S.A.,* 79, 3408, 1982.

48. **Fesenko, E. E., Kolesnikov, S. S., and Lyubarsky, A. L.,** Induction by cyclic-GMP of cationic conductance in the plasma membrane of retinal rod outer segment, *Nature,* 313, 310, 1985.

49. **Calhoon, R., Tsuda, M., and Ebrey, G.,** A light-activated GTP-ase from octopus photoreceptors, *Biochem. Biophys, Res. Commun.,* 94, 1452, 1980.

50. **Paulsen, R. and Bentrop, J.,** Light-modulated events in fly photoreceptors, *Prog. Zool.,* 33, 299, 1986.

51. **Blumenfeld, A., Erusalimsky, J., Heichal, O., Selinger, Z., and Minke, B.,** Light-activated guanosine triphosphates in *Musca* eye membranes resembles the prolonged depolarizing afterpotential in photoreceptor cells, *Proc. Natl. Acad. Sci. U.S.A.,* 82, 7116, 1986.

52. **Sitarayama, A. and Liebman, P. A.,** Phosphorylation of rhodopsin and quenching of cyclic GMP phosphodiesterase. Activation by ATP and weak bleaches, *J. Biol. Chem.,* 258, 12106, 1983.

53. **Kuhn, H.,** Proteins, involved in the control of cGMP phosphodiesterase in retinal rod cells, *Prog. Zool.,* 33, 287, 1986.

54. **Vandenberg, C. A. and Montal, M.,** Light-regulated events in invertebrate photoreceptors. II. Light-regulated phosphorylation of rhodopsin and phosphoinositides in squid photoreceptor membranes, *Biochemistry,* 23, 2347, 1984.

55. **Paulsen, R. and Bentrop, J.,** Reversible phosphorylation of opsin induced by irradiation of blowfly retinae, *J. Comp. Physiol. A,* 155, 39, 1984.

56. **Lisman, J. E.,** The role of metarhodopsin in the generation of spontaneous quantum bumps in UV receptors of *Limulus* median eye, *J. Gen. Physiol.,* 85, 171, 1985.

57. **Minke, B.,** Photopigment-dependent adaptation in invertebrates — implications for veterbrates, in *The Molecular Mechanism of Photoreception,* Dahlem Konferenzen 1986, Stieve, H., Ed., Springer-Verlag, Berlin, 1986, 241.

58. **Devary, O., Heichal, O., Blumenfeld, A., Cassel, A., Suss, A., Barash, A., Rubinstein, T., Minke, B., and Selinger, Z.,** Coupling of photoexcited rhodopsin to phosphoinositide hydrolysis in fly photoreceptors, *Proc. Natl. Acad. Sci. U.S.A.,* 84, 6939, 1987.

59. **Robinson, P. R. and Cote, R. H.,** Characterization of guanylate cyclase in squid photoreceptors, in preparation.

60. **Malbon, C. C., Kaupp, U. B., and Brown, J. E.,** *Limulus* ventral photoreceptors contain a homologue of the α subunit of mammalian N_s, *FEBS Lett.,* 172, 91, 1984.

61. **Tsuda, M., Tsuda, T., Terayama, Y., Fukuda, Y., Akino, T., Yamanaka, G., Stryer, L., Katada, T., Ui, M., and Ebrey, T.,** Kinship of cephalopod photoreceptor G-protein with vertebrate transducin, *FEBS Lett.,* 198, 5, 1986.

62. **Manning, D. R. and Gilman, A. G.,** The regulatory components of adenylate cyclase and transducin: a family of structurally homologous guanine nucleotide-binding proteins, *J. Biol. Chem.,* 258, 7059, 1983.

63. **Kanaho, Y., Su-Chen, T., Adamik, R., Hewlett, E. L., Moss, J., and Vaughan, N.,** Rhodopsin-enhanced GTP-ase activity of the inhibitory GTP-binding protein of adenylate cyclase, *J. Biol. Chem.,* 259, 7378, 1984.

64. **Stein, P. J., Halliday, K. R., and Rasenick, M. M.,** Photoreceptor GTP-binding protein mediates fluoride activation of phosphodiesterase, *J. Biol. Chem.,* 260, 9081, 1985.

65. **Fein, A. and Corson, D. W.,** Both photons and fluoride ions excite *Limulus* ventral photoreceptors, *Science,* 204, 77, 1979.

66. **Payne, R.,** Fluoride blocks an inactivation step of transduction in an insect photoreceptor, *J. Physiol. (London),* 325, 261, 1982.

67. **Minke, B. and Stephenson, R. S.,** The characteristics of chemically-induced noise in *Musca* photoreceptors, *J. Comp. Physiol.,* 156, 339, 1985.

68. **Yamanaka, G., Eckstein, F., and Stryer, L.,** Stereochemistry of the guanyl nucleotide binding site of transducin probed by phosphorothioate analogues of GTP and GDP, *Biochemistry,* 24, 8094, 1985.

69. **Steinweis, P. C., Northup, J. K., Smigel, M. D., and Gillman, A. G.,** The regulatory component of adenylate cyclase, *J. Biol. Chem.,* 256, 11517, 1981.

70. **Fein, A. and Corson, D. W.,** Excitation of *Limulus* photoreceptors by vanadate and a hydrolysis-resistant analogue of guanosine triphosphate, *Science,* 212, 555, 1981.

71. **Bolsover, S. R. and Brown, J. E.,** Injection of guanosine and adenosine nucleotides into *Limulus* ventral photoreceptor cells, *J. Physiol.,* 332, 325, 1982.

72. **Fein, A.,** Blockade of visual excitation and adaptation in *Limulus* photoreceptors by GDPβS, *Science,* 232, 1543, 1986.

73. **Eckstein, F., Cassel, D., Levkovitz, H., Lowe, M., and Selinger, Z.,** Guanosine 5'-O-(2-Thiodiphosphate), an inhibitor of adenylate cyclase stimulation by guanine nucleotides and fluoride ions, *J. Biol. Chem.,* 254, 9829, 1979.

74. **Walz, B., Coles, J. A., Poitry, S., and Levy, S.,** Light adaptation and the light-induced increase in intracellular $[Ca^{2+}]$ (and $[Na^+]$) are spatially localized in bee photoreceptors, *Verh. Dtsch. Zool. Ges.,* 79, 245, 1986.

75. **Conner, J. and Alkon, D. A.,** Light and voltage dependent increases of calcium ion concentration in molluscan photoreceptors, *J. Physiol. (London).,* 267, 299, 1977.

76. **Brown, J. E., Brown, P. K., and Pinto, L. H.,** Detection of light-induced changes of intracellular ionized clacium concentration in *Limulus* ventral photoreceptors using Arsenazo III, *J. Physiol. (London),* 267, 299, 1977.

77. **Levy, S. and Fein, A.,** Relationship between light-sensitivity and intracellular free calcium in *Limulus* ventral photoreceptors, *J. Gen. Physiol.,* 85, 805, 1985.

78. **Nagy, K. and Stieve, H.,** Changes in intracellular calcium ion concentration in the course of dark-adaptation measured by arsenazo III in the *Limulus* photoreceptor, *Biophys. Struct. Mech.,* 9, 207, 1983.

79. **Calman, B. G. and Chamberlain, S. C.,** Distinct lobes of *Limulus* photoreceptors. II. Structure and ultrastructure, *J. Gen. Physiol.,* 80, 839, 1982.

80. **Stern, J., Chinn, K., Bacigaloupo, J., and Lisman, J. E.,** Distinct lobes of *Limulus* ventral photoreceptors. I. Functional and anatomical properties of lobes revealed by removal of glial cells, *J. Gen. Physiol.,* 80, 825, 1982.

81. **Harary, H. H. and Brown, J. E.,** Spatially nonuniform changes in intracellular calcium ion concentration, *Science,* 225, 292, 1984.

82. **Payne, R. and Fein, A.,** Inositol 1,4,5 trisphosphate releases calcium from specialized sites within *Limulus photoreceptors, J. Cell Biol.,* 104, 933, 1987.

83. **Fein, A. and Payne, R.,** Specialization of the rhabdomeral lobe of *Limulus* ventral photoreceptors for phototransduction, in *Facets of vision: From Exner to Autrum,* Hardie, R. C. and Stavenga, D. G., Eds., Springer-Verlag, Berlin, in press.

84. **Bolsover, S. R. and Brown, J. E.,** Calcium, an intracellular messenger of light adaptation also participates in excitation of *Limulus* ventral photoreceptors, *J. Physiol. (London),* 364, 381, 1985.

85. **Clark, A. W., Millecchia, R., and Mauro, A.,** The ventral photoreceptors of *Limulus*. I. The microanatomy, *J. Gen. Physiol.,* 54, 289, 1969.

86. **Whittle, A. C.,** Reticular specializations in photoreceptors, a review, *Zool. Scr.,* 5, 191, 1976.

87. **Walz, B. and Fein, A.,** Evidence for calcium-sequestering smooth ER in *Limulus* ventral photoreceptors, *Invest. Ophthalmol. Vis. Sci. Suppl.,* 24, 281, 1983.

88. **Walz, B.,** Ca^{2+}-sequestering smooth endoplasmic reticulum in an invertebrate photoreceptor. I. Intracellular topography as revealed by OsFeCN staining *in situ* Ca accumulation, *J. Cell Biol.,* 93, 839, 1982.

89. **Walz, B.,** Ca^{2+}-sequestering smooth endoplasmic reticulum in retinula cells of the blowfly, *J. Ultrastruct. Res.,* 81, 240, 1982.

90. **Lisman, J. E. and Strong, J. A.,** The initiation of excitation and light adaptation in *Limulus* ventral photoreceptors, *J. Gen. Physiol.,* 73, 219, 1979.

91. **Berridge, M. J. and Irvine, R. F.,** Inositol 1,4,5 trisphosphate, a novel second messenger in cellular signal transduction, *Nature,* 312, 315, 1984.

92. **Hagins, W. A., Zonana, H. V., and Adams, R. G.,** Local membrane current in the outer segments of squid photoreceptors, *Nature (London),* 194, 844, 1962.

93. **Lasansky, A. and Fuortes, M. G. F.,** The site of origin of electrical responses in visual cells of the leech *Hirudo medicinalis, J. Cell Biol.,* 42, 241, 1969.

94. **Mauro, A.,** The ventral photoreceptor cells of *Limulus*. III. A voltage-clamp study, *J. Gen. Physiol.,* 54, 331, 1969.

95. **Millecchia, R. and Fein, A.,** Localization of the photocurrent of *Limulus* ventral photoreceptors using a vibrating probe, *Biophys. J.,* 50, 193, 1986.

96. **Brown, J. E. and Mote, M. I.,** Ionic dependence of reversal voltage of the light response in *Limulus* ventral photoreceptors, *J. Gen. Physiol.,* 63, 337, 1974.

97. **Bacigalupo, J. and Lisman, J. E.,** Ion channels activated by light in *Limulus*, ventral photoreceptors, *Nature,* 304, 268, 1983.

98. **Shimomura, O., Johnson, F. H., and Saiga, Y.,** Extraction, purification and properties of aequorin, a bioluminescent protein from the luminous hydromedusan *Aequorea, J. Cell. Comp. Physiol.,* 59, 223, 1962.

99. **Lisman, J. E. and Brown, J. E.,** The effects of intracellular iontophoretic injection of calcium and sodium ions on the light-response of *Limulus* ventral photoreceptors, *J. Gen. Physiol.,* 59, 701, 1972.

100. **Lisman, J. E. and Brown, J. E.,** Effects of intracellular injection of calcium buffers on light adaptation in *Limulus* ventral photoreceptors, *J. Gen. Physiol.,* 66, 489, 1975.

101. **Payne, R., Corson, D. W., and Fein, A.,** Pressure injection of calcium both excites and adapts *Limulus* ventral photoreceptors, *J. Gen. Physiol.,* 88, 107, 1986.

102. **Saibil, H. R.,** A light-stimulated increase in cyclic GMP in squid photoreceptors, *FEBS Lett.,* 168, 213, 1984.

103. **Johnson, E. C., Robinson, P. R., and Lisman, J. E.,** cGMP is involved in the excitation of invertebrate photoreceptors, *Nature,* 324, 468, 1986.

104. **Brown, J. E., Rubin, L. J., Ghalayini, A. J., Tarver, A. L., Irvine, R. F., Berridge, M. J., and Anderson, R. E.,** Myo-inositol polyphosphate may be a messenger for visual excitation in *Limulus* photoreceptors, *Nature (London),* 311, 160, 1984.

105. **Fein, A., Payne, R., Corson, D. W., Berridge, M. J., and Irvine, R. F.,** Photoreceptor excitation and adaptation by inositol 1,4,5 trisphosphate, *Nature (London),* 311, 157, 1984.

106. **Brown, J. E. and Rubin, L. J.,** A direct demonstration that inositol trisphosphate induces an increase in intracellular calcium in *Limulus* photoreceptors, *Biochem. Biophys. Res. Commun.,* 125, 1137, 1984.

107. **Payne, R., Corson, D. W., Fein, A., and Berridge, M. J.,** Excitation and adaptation of *Limulus* ventral photoreceptors by inositol 1,4,5 trisphosphate result from a rise in intracellular calcium, *J. Gen. Physiol.,* 88, 127, 1986.

108. **Corson, D. W. and Fein, A.,** Inositol 1,4,5, trisphosphate induces bursts of calcium release inside *Limulus* ventral photoreceptors, *Biophys. J.,* 47, 38a, 1985.

109. **Rubin, L. J. and Brown, J. E.,** Intracellular injection of calcium buffers blocks IP_3-induced but not light-induced electrical responses of *Limulus* ventral photoreceptors, *Biophys. J.,* 47, 380, 1985.

110. **Payne, R. and Fein, A.,** Rapid desensitization terminates the response of *Limulus* photoreceptors to brief injections of inositol trisphosphate (abstr.), *Biol. Bull.,* in press.

111. **Szuts, E. Z., Wood, S. F., Reid, M. A., and Fein, A.,** Light stimulates the rapid formation of inositol trisphosphate in squid retinae, *Biochem. J.,* 240, 929, 1986.

112. **Brown, J. E., Watkins, D. C., and Malbon, C. C.,** Light-induced changes of inositol phosphates in squid (*Loligo pealei*) retina, *Biochem. J.,* 247, 293, 1987.

113. **Wood, S. F., Szuts, E. Z., and Fein, A.,** Light-induced changes in inositol trisphosphate in distal segments of squid photoreceptors, *Invest. Opthalmol. Vis. Sci.,* 28(Suppl.), 96, 1987.

114. **Yoshioka, T., Takagi, M., Hayashi, F. and Amakawa, T.,** The effect of isobutylmethylxanthine on the photoresponse and phosphorylation of phosphatidylinositol in squid photoreceptor membranes, *Biochem. Biophys. Res. Commun.,* 755, 50, 1982.

115. **Baer, K. M. and Saibil, H. R.,** Light and GTP-activated hydrolysis of phosphatidylinositolbisphosphate in squid photoreceptor membranes, *J. Biol. Chem.,* in press

116. **Storey, D. J., Shears, S. B., Kirk, C. J. and Michell, R. H.,** Stepwise enzymatic dephosphorylation of inositol 1,4,5 trisphosphate to inositol in liver, *Nature,* 312, 374, 1984.

117. **Downes, C. P., Mussat, M. C., and Michell, R. H.,** The inositol trisphosphate monoesterase of the human erythrocyte membrane, *Biochem. J.,* 203, 169, 1982.

118. **Irvine, R. F., Anderson, R. E., Rubin, L. J., and Brown, J. E.,** Inositol 1,3,4 trisphosphate concentration is changed by illumination of *Limulus* ventral photoreceptors, *Biophys. J.,* 47, 38a, 1985.

119. **Irvine, R. F., Letcher, A. J., Heslop, J. P., and Berridge, M. J.,** The inositol tris/tetrakisphosphate pathway — a demonstration of Ins(1,4,5)P_3 3-kinase activity in animal tissues, *Nature (London),* 320, 631, 1986.

120. **Hansen, C. A., Mah, S., and Williamson, J. R.,** Formation and metabolism of inositol 1,3,4,5 tetrakisphosphate in liver, *J. Biol. Chem.,* 261, 8100, 1986.

121. **Wood, S. F., Szuts, E. Z., and Fein, A.,** Aluminum fluoride and GTP increase inositol phosphate production in distal segments of squid photoreceptors, *Biol. Bull.,* in press.

122. **Nathans, J. and Hogness, D. S.,** Isolation, sequence analysis, and inton-exon arrangement of the gene encoding human rhosopsin, *Proc. Natl. Acad. Sci. U.S.A.,* 81, 4851, 1983.

123. **Nathans, J., Thomas, D., and Hogness, D. S.,** Molecular genetics of human color vision: the genes encoding blue, green and red pigments, *Science,* 232, 193, 1986.

Index

INDEX

A

Adenylate cyclase, 53, 60
  growth factor receptors and, 30
  inhibition of, pertussis toxin and, 4—5
  olfactory transduction and, 128—130
  ras proteins and, in yeast, 37
ADP-ribosylation, 5, 141
  astrocytoma cells and, 61
  cholera toxin, 127
  G_{PLC} identification and, 69
  pertussis toxin-catalyzed, G proteins as substrate of, 7—11
Adsorption model for olfactory reception, 127
Aequorin, 145
Alpha$_1$-adrenergic receptors 50, 52
AMP, cyclic, see Cyclic AMP
Angiotensin II, phospholipase A$_2$ and, 52
A-protomer, 6
Astrocytoma cells, and muscarinic receptor-stimulated PI hydrolysis, 60—61

B

Bacterial toxins, G_{PLC} function regulation by, 63—65
Beta-gamma subunits, 53
B-oligomer, 6
Bordetella pertussis, 4
Botulinum toxin, G_{PLC} identification and, 69
Bradykinin, phospholipase A$_2$ and, 52

C

Ca^{2+}, cytosolic, 96
Ca^{2+} currents, 80—91
  cyclic AMP-independent hormonal modulations of, 82, 84—85
  G proteins involved in modulations of, 85—89
Ca^{2+} mobilizing receptors, G proteins coupling to, 61—63
c-*abl* oncogene, 33
Calcium channel, regulation of, pertussis toxin-susceptible, 13—14
Calcium movements
  GTP mechanism of action on, 106—115
    calcium uptake and, 107—110
    clues to, 109, 111
    GTP-induced loading of IP$_3$-releasable pool and, 113—115
    transmembrane calcium "conveyance" model, 111—113
  inositol phosphate-activated, 97, 103—106
Calcium-regulatory organelles, 96—97
Calcium release
  guanine nucleotide-activated, 98—103
    IP$_3$-activated release and, 103—106
    IP$_3$-mediated, 103—106, 145—147
    by light, in photoreceptors, 136—149, see also

Photoreceptors
Calcium translocation, mechanism of, GTP and, 101—103
Calcium uptake, GTP-activated, 107—109
cAMP, see Cyclic AMP
Carbachol, PI hydrolysis and, 61
cDNA clones, 127, 128
Cell proliferation and differentiation, pertussis toxin-susceptible, 14—17
c-*erb*B1 oncogene, 31
c-*erb*B2 oncogene, 31—32
c-*fms* oncogene, 32
cGMP, messenger, 145
Cholera toxin, 7, 53, 60, 65, 140
Cholera toxin ADP-ribosylation, 127
Cilia, olfactory neurons and, 125
c-*neu* oncogene, 31—32
Counterport process, 48
CPAE cells, phospholipase A$_2$ and, 50—52
c-*src* oncogene, 32
Cyclic AMP, 8, 84—85
Cytoskeleton, 138
Cytosolic Ca^{2+}, 96

D

Diacylglycerol, 60
Drosophila, photoreceptors of, 136, see also Rhodopsin

E

Effector region of *ras* proteins, 34
EGF (epidermal growth factor) receptor, 31
Eicosanoids, 48
Endoplasmic reticulum
  inositol phosphate-activated calcium movements and, 97
  smooth, and calcium release by light, 143, 149
Epidermal growth factor (EGF) receptor, 31
Epinephrine, hyperglycemia induced by, pertussis vaccine in abolition of, 4

F

Fluoride
  PI hydrolysis and, 62, 65
  transducin activation and, 142
f-Met-Leu-Phe, in neutrophils, 49
FRTL5 cells, phospholipase A$_2$ and, 50

G

G_s, 7
  adenylate cyclase and, 60
  $\alpha-$ subunit release from, 8
G_i, 8, 60
G_i subfamily, 7—9
G_o, 7, 8

G_{PLC}
  identification of, 67—69
  regulation of, 63—67
G_T, 7, 8
GAP (GTPase activating protein), 35—36, 38
GDP, 9
Genes, 38, see also Oncogenes
Glucocorticoids, 48, 52
G proteins, diversity of, 7—9
Growth factor receptor oncogenes, 30—32
GTP
  calcium uptake activated by, 107—109
  exchange of, with prebound GDP, 9
  hydrolysis of, to GDP, 9
  mechanism of action of, 116—117
    on calcium movements, 106—115
    IP_3 mechanism of action and, 103—105
  reversibility of action of, 101
  transmembrane calcium "conveyance" model
    activated by, 111—113
GTP analogs, light-activated binding of, to microvillar membrane, 140
GTPase activating protein (GAP), 35—36, 38
GTPase activity, 60, 140
GTP photoaffinity labels, 141—142
Guanine nucleotide analogs, and Ca^{2+} current
  modulations, 85—86
Guanine nucleotides
  calcium release activated by, 98—103
  cellular and subcellular specificity of,
    100—101
  sensitivity and specificity of, 98—100

H

Hormones, Ca^{2+} current modulations by, cyclic AMP-
  independent, 82, 84—85
Hydroxyeicosatetraenoic acids, 48
Hyperglycemia, epinephrine-induced, pertussis
  vaccine in abolition of, 4

I

IAP (islet-activating protein), 4
IAP (islet-activating protein) substrates, 127
Immunoblotting analysis, 127
Immunohistochemical analysis, 127
Inositol phosphate, calcium movements activated by,
  97, 103, 106, 145—147
Inositol 1,4,5-trisophosphate, 60
  calcium pool releasable by, GTP-induced loading
    of, 113—115
  light-induced production of, evidence for,
    147—148
  mechanism of action of, 103—105, 116—117
Intracellular calcium signaling, 96—97
Intracellular tyrosine kinases, 32—33
Invertebrates, photoreceptors of, 136, see also
  Photoreceptors
Islet-activating protein (IAP), 4
Islet-activating protein (IAP) substrates, 127

L

Lectin inhibition of receptor binding, olfactory
  receptors and, 126
Leukotriene D_4, CPAE cells and, 50
Leukotrienes, 48
Ligand binding studies, of olfactory receptors, 126
Light, calcium release by, in photoreceptors, 136—
  149, see also Photoreceptors
Light-activated G proteins, evidence for, in
  microvillar photoreceptors, 139—140
Limulus photoreceptors, 142, 143, 146
Lipocortin, 48

M

Macrophages, RAW264.7, 53
Mast cells, 49
MDCK cells, phospholipase A_2 and, 52
Membrane interactions, GTP and, 101—103
Messenger cGMP, 145
Metarhodopsin, invertebrate, phosphorylation of, 140
Metoprolol, FRTL5 cells and, 50
Microvillar membrane(s)
  identification of G proteins in, 140—142
  light-activated binding of GTP analogs to, 140
Microvilli, 137—138, see also Photoreceptors,
  microvillar
Muscarinic receptors, 60—70
  G_{PLC} identification and, 67—69
  G_{PLC} regulation and, 63—67
  phospholipase C activation and, 60—63

N

Neomycin, phospholipase C and, 50
Neurons, olfactory, 124—132, see also Olfactory
  entries
Neutrophils, f-Met-Leu-Phe in, 49
N-*ras* protooncogene, 38

O

Olfactory neurons, 124—132
Olfactory receptors, 125—128
Olfactory transduction, G proteins in, 127—131
Oncogenes
  growth factor receptor, 30—32
  intracellular tyrosine kinases, 32—33
  ras proteins, 33—38
  signal transduction and, 30—39
Opsin, 136, 137
Organelles, calcium-regulatory, 96—97
Oxalate-precipitation of calcium, GTP-activated
  calcium uptake and, 107—108

P

Palmitic acid, 34
PDGF receptor, 32
Pertussis toxin, 60, 140

A-B structure of, 5—6
ADP-ribosylation catalyzed by, G proteins as
  substrate of, 7—11
Ca^{2+} current modulations and, 86—87
identification of, 4
as probe of G proteins, 4—6
Pertussis toxin substrate G proteins, 4—18
  mediating signaling other than adenylate cyclase
    system, 11—17
    calcium channel regulation, 13—14, 86—87
    cell proliferation and differentiation, 14—17
    phospholipase A_2 activation, 12—13, 49—51
    phospholipase C activation, 11—12
    potassium channel activation, 13
Pertussis vaccine, epinephrine-induced hyperglycemia abolition by, 4
Phorbol esters
  PI hydrolysis and, 66
  protein kinase C and, 67
  in 3T3 cells, phospholipase A_2 and, 52
Phosphatidylinositol, neomycin and, 50
Phosphatidylinositol 1,4 biphosphate, 148
Phosphoinositide hydrolysis
  muscarinic receptor-stimulated, 60—61
  olfactory transduction and, 130—131
  phorbol esters and, 66
Phosphoinositides, 60
Phospholipase A_2
  activation of, pertussis toxin-susceptible, 12—13
  coupled to inhibitory G proteins, 53
  G-protein regulation of, 48—56
    CPAE cells and, 52
    human platelets and, 52
    MDCK cells and, 52
    mechanism for, 53
    pertussis toxin-insensitive G protein and, 51—52
    pertussis toxin-sensitive G protein and, 50—51
    rabbit kidney proximal tubule cells and, 52
  v-*src* oncogene and, 33
Phospholipase A_2-activating protein (PLAP), 48
Phospholipase C, 48, 50, 51, 60, 70, 148
  activation of
    G protein involvement in, 60—63
    pertussis toxin-susceptible, 11—12
    ras proteins and, 37
  control of, by rhodopsin, 148—149
  in CPAE cells, 52
  phosphorylation of, 66
Phosphorylation
  G_{PLC} function regulation by, 65—67
  of invertebrate metarhodopsin, 140
  of opsin, 137
Photoreceptors, 136—149, see also Rhodopsin
  microvillar, 142—149
    calcium ions' role in light response and, 144—
      145
    calcium release from internal stores by light and,
      143—144
    evidence for light-activated G proteins in, 139—
      140
    inositol 1,4,5 triphosphate-mediated calcium

release and, 145—147
    light-induced production of inositol 1,4,5
      triphosphate and, 147—148
    phospholipase C control by rhodopsin and, 148—
      149
    physiological evidence for G protein involvement
      in phototransduction and, 142—143
Phototransduction, 136, 142—143
PI kinase, 33
PLAP (phospholipase A_2-activating protein), 48
Platelets, human, phospholipase A_2 and, 52
Potassium channel, activation of, pertussis toxin-
  susceptible, 13
Prazosin, FRTL5 cells and, 50
Prostaglandins, 48
Protein kinase C, 48, 60
  G_{PLC} function and, 65
  G_{PLC} identification and, 69
  phorbol esters and, 67
Protooncogenes, 30, 38, see also Oncogenes

R

ral gene, 38
ras proteins, 30, 33—38
  biochemical activity of, 34—36
  function of, 36—38
  genes related to, 38
  and G_{PLC} identification, 68—69
  mutational analysis of, 36
  structure of, 33—34
RAW264.7 macrophages, 53
Receptor molecules, rhodopsin and, G protein
  activation and, 138—142
Receptors
  Ca^{2+}-mobilizing, G protein coupling to, 61—63
  G protein uncoupling from, after ADP-ribosylation
    by pertussis toxin, 10—11
  growth factor, oncogenes and, 30—32
  muscarinic, 60—70
  olfactory
    characteristics of, 125—127
    G-protein coupling to, 128
Retinal rod outer segments, phospholipase A_2 in,
  51, 53
Rhodopsin, 136—138, see also Photoreceptors
  G protein activation and, 138—142
  phospholipase C control by, 148—149
rho genes, 38
R-*ras* gene, 38

S

Signal transduction, 30—39, see also Transducers
Smooth endoplasmic reticulum, and calcium release
  by light, 143, 149

T

3T3 cells, 49, 52
Thromboxanes, 48

Thyroid cells, phospholipase A$_2$ and, 50
Transducers, 9—10, see also Olfactory transduction;
    Signal transduction
Transducin
    light and, 139
    in retinal rod outer segments, phospholipase A$_2$ and,
        51, 53
    rhodopsins and, 138
Transmembrane calcium "conveyance" model, GTP-
    activated, 111—113
Tumor necrosis factor, phospholipase A$_2$ and, in

CPAE cells, 52
Tyrosine kinase(s), 31—33

V

v-*abl* oncogene, 33
Vertebrates, photoreceptors of, 136
v-*src* oncogene, 32, 33

Y

Yohimbine, FRTL5 cells and, 50